Progress in Clinical Parasitology
Volume III

Tsieh Sun, M.D.
Editor-in-Chief

Progress in Clinical Parasitology
Volume III

With 70 Illustrations

Springer-Verlag
New York Berlin Heidelberg London Paris
Tokyo Hong Kong Barcelona Budapest

Tsieh Sun, M.D.
Professor and Chief, Division of Clinical Pathology
North Shore University Hospital
Cornell University Medical College
300 Community Drive
Manhasset, NY 11030
USA

ISSN: 1062-0338

Printed on acid-free paper.

© 1993 Springer-Verlag New York, Inc.
Softcover reprint of the hardcover 1st edition 1993
Volumes I and II of this title were published by Field and Wood, Inc.
All rights reserved. This work may not be translated or copied in whole or in part without the written permission of the publisher (Springer-Verlag New York, Inc., 175 Fifth Avenue, New York, NY 10010, USA), except for brief excerpts in connection with reviews or scholarly analysis. Use in connection with any form of information storage and retrieval, electronic adaptation, computer software, or by similar or dissimilar methodology now known or hereafter developed is forbidden.
The use of general descriptive names, trade names, trademarks, etc., in this publication, even if the former are not especially identified, is not to be taken as a sign that such names, as understood by the Trade Marks and Merchandise Marks Act, may accordingly be used freely by anyone.
While the advice and information in this book is believed to be true and accurate at the date of going to press, neither the authors nor the editors nor the publisher can accept any legal responsibility for any errors or omissions that may be made. The publisher makes no warranty, express or implied, with respect to the material contained herein.

Production managed by Karen Phillips; manufacturing supervised by Jacqui Ashri.
Typeset by Asco Trade Typesetting Ltd, Hong Kong.

9 8 7 6 5 4 3 2 1

ISBN-13: 978-1-4612-7646-3   e-ISBN-13: 978-1-4612-2732-8
DOI: 10.1007/978-1-4612-2732-8

Preface

This volume, now the third in a series, presents a more heterogeneous content than previous issues. It covers two previously rare but now common opportunistic infections in the United States, a common parasitic disease in Japan, exciting but difficult problems in developing a malarial vaccine, a study exemplifying the role of T-lymphocytes in parasitic infections, and a fascinating review of the relationship between the schistosomes and their molluscan hosts.

The first chapter covers cryptosporidiosis, which has become a household name since the outbreak of the acquired immunodeficiency syndrome (AIDS). However, infection is now recognized to occur widely in immunocompetent individuals, with clustering of infection among veterinary students, laboratory workers, children in day care centers, and family members. It can also be the cause of traveler's diarrhea and nosocomial infection. Indeed, *Cryptosporidium* has become recognized as the leading protozoal cause of diarrhea worldwide. This chapter provides a concise, yet comprehensive, review on aspects of epidemiology, microbiology, clinical features, diagnosis, and treatment of this important disease. Recent in vitro studies of *Cryptosporidium*, conducted in Dr. Flanigan's and other laboratories, are described. They complement the extensive clincial experience of Dr. Soave, who summarizes her many articles in this field.

The second chapter describes another common opportunistic infection among AIDS patients, toxoplasmosis. This disease differs from cryptosporidiosis in that it was recognized as a common infection in immunocompetent individuals even before the AIDS outbreak. It can become a devastating accompaniment of AIDS, however, as it always involves the brain; and it has become the most common cause of a space-occupying cerebral lesion in AIDS victims. Drs. Decker and Tuazon, who work in a hospital where many patients with toxoplasmic encephalitis are diagnosed and treated, present their experience in this chapter. Also included is some interesting information concerning immunotherapy, particularly derived from the detection of a 23-KD (p23), the major antigen secreted by *Toxoplasma gondii*, which may become an essential ingredient in the production of *Toxoplasma* vaccine.

Chapter 3 describes a parasitic disease that is not yet a familiar name in the United States. Anisakidosis, previously known as anisakiasis, is common in Japan; up to 1990 there had been 16,090 cases reported in this country compared to 559 cases reported elsewhere. It is one of two human parasitic infections that is transmitted by saltwater fish. With an increasing number of Americans developing a taste for raw fish dishes, such as "sushi," "sashimi," and "ceviche," this disease may become more prevalent here in the future. Dr. Hajime Ishikura and his associates are indisputable experts in this field. Dr. Ishikura has edited two authoritative monographs on anisakidosis, published in 1990. This chapter not only covers such general topics as epidemiology, parasitology, pathology, and clinical features, it presents state-of-the-art knowledge regarding the diagnosis of this disease by seroimmunologic methods and molecular biologic techniques. In fact, several new methods were developed by this group of experts.

The topic of Chapter 4 is malarial vaccine. Because malaria is associated with high rates of morbidity and mortality in developing countries, vaccine production is undoubtedly of utmost importance. Dr. Hoffman and his team work in one of the few medical centers in the United States engaged in this difficult battle. This chapter covers the history of malarial vaccine development and the principles and theories of vaccine production. Obviously, the most appropriate target to attack is the sporozoite, which is the infective stage of the parasite. With the advent of molecular biology, specific synthetic or recombinant subunit molecules can be used to induce a potent immunity to sporozoites. After reading this chapter, one appreciates the tantalizing possibility of producing a useful malarial vaccine and the difficult problems that remain.

It has been known for some time that the human T-cell plays an important role in immunity against parasitic infections. In Chapter 5, Dr. Reed and his team describe T-cell functions and the relation of this cell to the clinical process of leishmaniasis. Recent basic advances in T-cell physiology permit a deeper insight into this relation than was possible previously. For instance, the authors studied the newly identified T-cell subset that carries the γ/δ T-cell receptor. Their studies show that T-cell cytokine production is closely related to delayed hypersensitivity; both IL-2 and IFN-T are negative during acute visceral leishmaniasis but become positive upon recovery. Healing cutaneous leishmaniasis is characterized by strong T-cell responses, and diffuse cutaneous leishmaniasis by weak T-cell responses. These studies may provide the basis for cytokine therapy for leishmaniasis, and experience with this disease may prove useful for other infections.

The topic of the last chapter may not be familiar to some readers. However, this significant basic study is highly regarded in the field of malacology, and the author, Dr. George Davis, is one of the few internationally known malacologists. The study is valuable especially because most data were collected in difficult situations, such as during the Vietnam War and inside communist countries. The coevolution of a parasite and its snail host is based on the fact that a mutation that protects the snail from infection is countered by a mutation in the parasite to overcome the snail's defense. On the other hand, ancient environmental changes, such as the development of the Himalayan mountain range, can have had important influences on the migration and spreading of snail populations, as is discussed under the fascinating section Historic Reconstruction. Dr. Davis' theory has led to the discovery of several new species of snails and schistosomes.

* * * *

This volume is dedicated to a legendary figure in tropical medicine, Philip Davis Marsden, O.B.E., M.D. (London), F.R.C.P., F.R.C.P.E., D.T.M. & H. (England), D.A.P. & E. (London). Dr. Marsden was trained at the London School of Hygiene and Tropical Medicine. His experience in tropical medicine is vast, as outside his native land, he served in many countries throughout the world, including Nigeria, Ghana, Gambia, Uganda, Tanzania, Kenya, New Guinea, Brazil, and the United States. Dr. Marsden has been Professor of Medicine at the University of Brasilia for the last 17 years; in addition, he currently holds honorary appointments at Cornell University Medical College, New York; the Uniformed Services University of the Health Sciences, Bethesda; and London School of Hygiene and Tropical Medicine.

His research interest is insect-borne protozoal disease, and his recent research has been mainly on the control of Chagas' disease and clinical aspects of *Leishmania viannia brasiliensis* infection. In Gambia, West Africa, he conducted one of the first longitudinal studies of tropical child health, which led to his M.D. thesis. In Uganda he led the team that defined "hyperimmune malarious splenomegaly."

However, Dr. Marsden's interest covers the broad spectrum of tropical medicine: He has published nearly 300 scientific articles and book chapters, and he has been a consultant editor and contributor to the parasitology section in *Cecil's Textbook of Medicine* for editions 14 through 16. However, I am most impressed with his recent series of letters on tropical medicine published in the *British Medical Journal*. When I read them, the image of his vivid lecturing style comes to mind immediately. Students have always thronged to his lectures, atrracted as much by his enthusiasm and wit as by his erudition. Dr. Marsden is indeed a great physician, researcher, and teacher. We therefore take this opportunity to acknowledge his outstanding contribution in the field of tropical medicine.

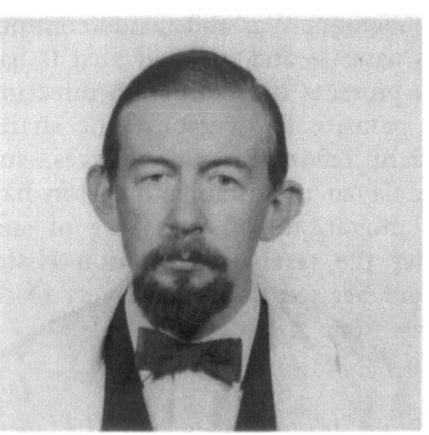

Dr. Philip Davis Marsden

Contents

Preface v
Editorial Board xi
Contributors xiii

1 Cryptosporidiosis 1
  Timothy P. Flanigan and Rosemary Soave

2 Toxoplasmosis: An Update on Clinical and
  Therapeutic Aspects 21
  Catherine F. Decker and Carmelita U. Tuazon

3 Anisakidae and Anisakidosis 43
  Hajime Ishikura, Kokichi Kikuchi, Kazuya Nagasawa,
  Toshio Ooiwa, Hiroshi Takamiya, Noriyuki Sato, and
  Kazuo Sugane

4 Antibodies to the Circumsporozoite Protein and
  Protective Immunity to Malaria Sporozoites 103
  Trevor R. Jones, W. Ripley Ballou, and Stephen L. Hoffman

5 Human T-Cell Responses in *Leishmania* Infections 119
  Donna M. Russo, Manoel Barral-Netto, Aldina Barral, and
  Steven G. Reed

6 Evolution of Prosobranch Snails Transmitting Asian
  Schistosoma; Coevolution with *Schistosoma*: A Review 145
  George M. Davis

Index 205

Editorial Board

Editor-in-Chief
Tsieh Sun, M.D., Manhasset, New York USA

Board Members

Lawrence R. Ash, Ph.D., Los Angeles, California USA

John H. Cross, Ph.D. Bethesda, Maryland USA

George R. Healy, Ph.D., Atlanta, Georgia USA

Warren D. Johnson, M.D., New York, New York USA

Thomas C. Jones, M.D., Basel, Switzerland

B.H. Kean, M.D., New York, New York USA

Mao Shou-Pei, M.D., Shanghai, People's Republic of China

Philip Marsden, M.D., F.R.C.P., Brasilia, Brasil

Henry Masur, M.D., Bethesda, Maryland USA

Harry Most, M.D., New York, New York USA

Robert L. Owen, M.D., San Francisco, California USA

Moriyasu Tsuji, M.D., D.M.S., Hiroshima, Japan

Editorial Board

Contributors

W. Ripley Ballou, M.D., Chief, Department of Immunology, Walter Reed Army Institute of Research Washington, D.C. USA; Assistant Professor, Uniformed Services University of the Health Sciences, Bethesda, MD USA

Aldina Barral, Ph.D., Associate Professor of Pathology, Federal University of Bahia, School of Medicine, Rua Joao Das Botas X/N Canela, 40.140-Salvador-Bahia, Brazil

Manoel Barrel-Netto, Ph.D., Associate Professor of Pathology, Federal University of Bahia, School of Medicine, Rua Joao Das Botas X/N Canela, 40.140-Salvador-Bahia, Brazil

George Davis, Ph.D., Pilsbry Chair of Malacology, Chairman, Department of Malacology, The Academy of Natural Sciences, 19th and the Parkway, Philadelphia, PA 19103 USA

Catherine F. Decker, M.D., Assistant Professor of Medicine, Uniformed Services University of the Health Sciences, F. Edward Hebert School of Medicine, Bethesda, MD USA

Timothy Flanigan, M.D., Assistant Professor of Medicine, International Health Institute, Brown University, Box G-BMC, 497, Providence, 02912 USA

Stephen L. Hoffman, M.D., Director, Malaria Program Naval Medical Research Institute Bethesda, MD; Adjunct Associate Professor, Uniformed Services University of the Health Sciences, Bethesda, MD USA

Hajime Ishikura, M.D., Ph.D., Department of Pathology, Sapporo Medical College, Chuo-ku, 060 Sapporo, Japan

Trevor R. Jones, Ph.D., Deputy Director, Malaria Program, Naval Medical Research Institute, Bethesda, MD USA

Kokichi Kikuchi, M.D., Ph.D., President and Professor, Department of Pathology, Sapporo Medical College, Chuo-ku, 060 Sapporo, Japan

Kazuza Nagasawa, Ph.D., Research Scientist, National Research Institute of Far Seas Fisheries, Shimizu, Shizuoka, Japan

Toshio Ooiwa, M.D., Director, Ooiwa Gastrointestinal Hospital, Fukuoka, Japan

Steven G. Reed, Ph.D., Senior Scientist, Seattle Biomedical Research Institute, 4 Nickerson Street, Seattle, WA 98109; Adjunct Associate Professor of Medicine, Cornell University Medical College, New York, NY USA

Donna M. Russo, Ph.D., Postdoctoral Fellow, Seattle Biomedical Research Institute, 4 Nickerson Street, Seattle, WA 98109; Adjunct Research Assistant Professor of Medicine, Cornell University Medical College, New York, NY USA

Noriyuki Sato, M.D., Ph.D., Associate Professor, Department of Pathology, Sapporo Medical College, Chuo-ku, 060 Sapporo, Japan

Rosemary Soave, M.D., Associate Professor, Department of Medicine and Public Health, New York Hospital-Cornell University Medical College, New York, NY USA

Kazuo Sugane, M.D., Professor, Department of Parasitology, Faculty of Medicine, Shinshu University, Matsumoto, Japan

Hiroshi Takamiya, M.D., Director, Takamiya Hospital, Fukuoka, Japan

Carmelita U. Tuazon, M.D., Professor of Medicine, Director, Division of Infectious Diseases, George Washington University Medical Center, 2150 Pennsylvania Avenue, Washington, D.C. 20037 USA

1
Cryptosporidiosis

Timothy P. Flanigan and Rosemary Soave

Though *Cryptosporidium* was first described by E.E. Tyzzer in the gastric glands of laboratory mice in 1907 (1), it has only been during the last decade, with the advent of the epidemic of acquired immunodeficiency syndrome (AIDS), that this protozoan parasite has been identified as a significant and ubiquitous human pathogen. It is the cause of cryptosporidiosis, a newly recognized disease that is currently undergoing definition and characterization. *Cryptosporidium* has been described as both rare and common, a benign commensal organism and a malignant opportunistic pathogen responsible for life-threatening enteritis (2); in fact, the spectrum of illness caused by this enigmatic parasite is broad and closely linked to host immunocompetence.

Between 1907 and 1955 *Cryptosporidium* was described in many species of animals, including turkeys, rattlesnakes, and the European common fox (3–5), but it was thought to be nonpathogenic and thus received little attention. Over the next 20 years the association between cryptosporidial infection and diarrheal disease was made, and the resultant morbidity in domestic animals such as cows, horses, and pigs was appreciated (6–8). The first human case of cryptosporidiosis, reported in 1976, was that of a 3-year-old child in rural Tennessee who presented with severe gastroenteritis of 2 weeks' duration (9). Cryptosporidial infection was diagnosed by electron microscopic examination of a biopsy specimen of intestinal mucosa. The child was otherwise healthy and recovered spontaneously without sequelae. Fewer than a dozen cases of cryptosporidiosis were reported over the next 5 years (10). The spectrum of illness ranged from self-limited diarrhea in immunocompetent individuals, who for the most part were exposed to domestic animals, to often fatal enteritis in immunodeficient persons. Two children with congenital hypogammaglobulinemia and cryptosporidial infection suffered with unremitting diarrhea and malabsorption for more than 3 years, and it was unresponsive to all therapy (11,12).

In 1982 the Centers for Disease Control (CDC) reported 21 homosexual men with severe diarrhea and the presence of *Cryptosporidium* oocysts in stool samples (13). The severity of diarrhea was extraordinary (up to 15 liters of stool per day) and the mortality in this group was high, foreshadowing the morbidity that this protozoan infection was to cause in human immunodeficiency virus (HIV)-infected individuals over the next 10 years. Subsequent popularization of simple and inexpensive techniques for detecting oocysts in stool specimens has led to the recognition of *Cryptosporidium* as the leading protozoan cause of diarrhea in immunocompetent children and as a devastating opportunistic infection in immunodeficient persons.

Microbiology

The protozoan *Cryptosporidium*, which means hidden spore in Greek, has been placed within the class Sporozoa on the basis of the oocyst having four sporozoites (14) (Figure 1.1). It is thus taxonomically related to other pathogenic protozoa, including *Toxoplasma gondii* and *Isospora belli*. which cause opportunistic infections in patients with AIDS (14). Although more than 20 species of *Cryptosporidium* have been described and named for various animal species in which they were found, cross-transmission studies suggest that the parasite lacks host specificity (15). The differences between the various species and isolates are still not fully defined. *Cryptosporidium parvum* has been designated the species that causes disease in humans.

Unlike the related parasite *Toxoplasma*, the complete life cycle of *Cryptosporidium* is monoxenous; i.e., it occurs within a single host. Cryptosporidial infection in man is initiated by ingestion of the oocyst with subsequent excystation or release of four sporozoites

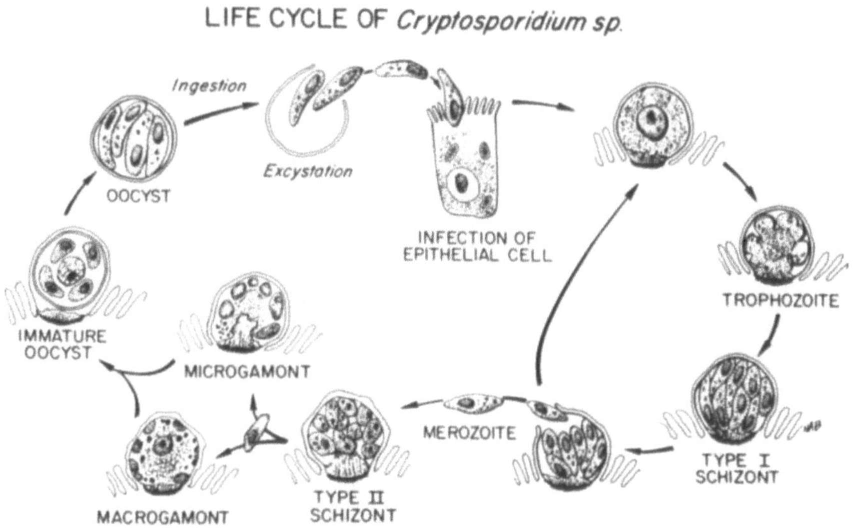

Figure 1.1. Life cycle of *Cryptosporidium* sp.

within the gastrointestinal tract (Figure 1.2). Although excystation may occur spontaneously, it appears to be more efficient in the presence of bile salts and digestive enzymes (16). Sporozoites implant immediately on the host epithelial cell and subsequently develop into the endogenous stages of the parasite at the luminal surface of the epithelium. *Cryptosporidium* has been identified within M cells of the gastrointestinal tract of symptomatic guinea pigs (17), but infection does not cause ulceration of the epithelial layer, nor is there clear evidence that it invades the lymphatic or systemic circulation. Small intestinal biopsy specimens from patients with AIDS and severe infection typically show a cellular infiltrate in the lamina propria, partial villous atrophy, and parasites present at the microvillous border of the epithelial cells with subsequent damage to the brush border (18) (Figure 1.3). By light microscopy the parasite appears to be extracellular, but on closer examination by electron microscopy it is seen to be intracellular, underneath the cell membrane, but in an "extracytoplasmic" position. The interface between the parasite and the cell surface is clearly delineated by a dense band and a feeder layer.

Upon implantation the sporozoite develops into a trophozoite, which buds to form eight distinct merozoites within a type I schizont. This process of asexual multiplication can occur within as little as 24 hours in cell culture, indicating rapid protein synthesis, which is further evidenced by abundant rough endoplasmic reticulum within the trophozoite (19). The merozoite is morphologically similar

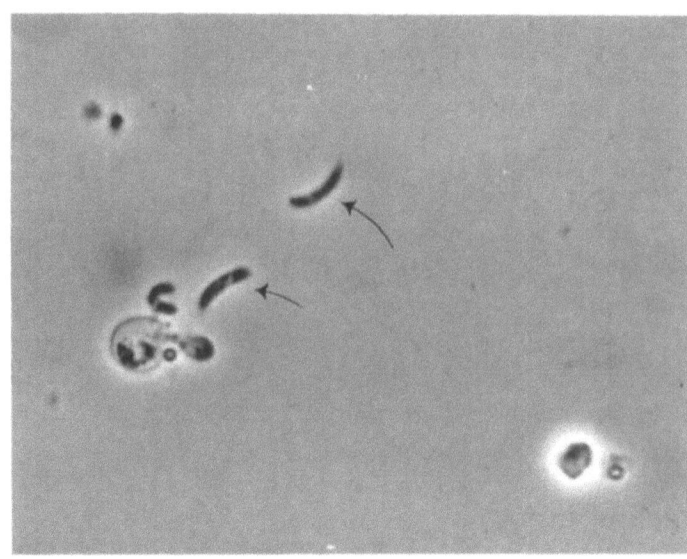

Figure 1.2. *Cryptosporidium* sporozoites (arrows) released upon oocyst excystation. (×630 phase contrast)

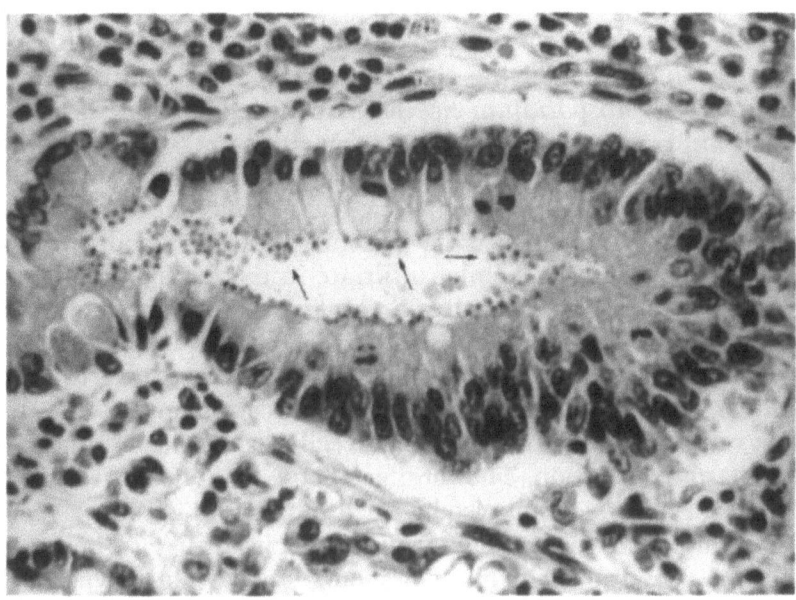

Figure 1.3. Duodenal biopsy sample from a patient with AIDS and cryptosporidiosis. Numerous parasites (arrows) are seen in the microvillous border. (hematoxylin and eosin)

to the sporozoite containing micronemes and rhoptery bodies. Type I schizonts release the merozoites, which can invade other epithelial cells and develop into either type I or type II schizonts. Type II schizonts may produce up to 16 merozoites, which in turn can reinitiate schizogony or enter into the sexual phase of the cycle and develop into micro- or macrogametocytes. The ability of the merozoite to continue to cycle within the asexual phase endows the organism with enormous potential for autoinfection. Clinically, this trait is reflected in the large parasite burden and continuous oocyst shedding observed in AIDS patients with cryptosporidiosis. Fertilization of micro- and macrogametocytes results in zygote formation and oocyst maturation. When released from the epithelial cell into the lumen, the oocyst is mature (sporulated); and either it is excreted into the environment where it is immediately infectious, or it excysts within the bowel lumen and reinfects the same host. The oocyst is surrounded by a carbohydrate-rich substance that stains avidly with ruthinium red and is consistent with a glycocalyx (20).

Using both light and electron microscopy, some investigators have described two distinct forms of the oocyst: a thin-walled oocyst and a thick-walled oocyst (16). Thin-walled oocysts are believed to be primarily infective to the same host, whereas the hardier, thick-walled oocysts are excreted into the environment and are thus primarily responsible for transmission of infection.

Epidemiology

Initially *Cryptosporidium* was thought to be pathogenic for immunocompromised persons and only a rare cause of disease in immunocompetent individuals. It is now recognized to be a leading protozoal cause of diarrhea worldwide and has been reported in six continents. Reports of infection have included persons of all ages, including a 3-day-old infant and a 95-year-old grandparent (21,22). The incidence of symptomatic enteritis appears to be greatest in infants and younger children, likely reflecting both increased fecal-oral transmission of pathogens and the absence of protective immunity. The true prevalence of cryptosporidiosis is not known. Data obtained from studies of children with diarrhea (excluding day care center outbreaks) revealed prevalence rates ranging from 1.4% to 22.2%. The mean prevalence rate from 38 studies conducted in various parts of the world was 7.6% (10).

The overall prevalence of infection in developing countries is reported to be 4.9% in Asia and 10.4% in Africa. In comparison, the prevalence rate is 1–3% in North America and Europe. Increased rates of infection appear to be related to lack of clean water, poor sanitation, crowding, and close proximity between humans and domestic animals. It is not surprising that cryptosporidiosis is a common cause of traveler's diarrhea in persons returning from a less developed country (23). In general, urban areas have a higher incidence of infection than do rural areas.

Seroprevalence rates in immunocompetent individuals in the United States are between 25% and 35%, whereas in Latin America they are well over 50%, indicating that infection is much more common than surveys based on fecal oocyst shedding suggest (24). In fact, in a Peruvian serologic survey, 73% of the children between the ages of 2 and 5 years old had detectable anticryptosporidial immunoglobulin G (IgG).

Breast-fed children appear to have fewer bouts of cryptosporidial diarrhea, which may be due to decreased exposure to infected water supplies and to specific protective factors such as anti-cryptosporidial IgA found in mother's milk (25,26). Seasonal differences in infection rates have been observed with an increase in infection during the warmer and more humid months.

Cryptosporidium is a common cause of diarrhea in patients with AIDS in the United States, but the true incidence is as yet undetermined. In two prospective studies at the National Institutes of Health and at the Johns Hopkin's Hospital, 15% and 16%, respectively, of patients with AIDS and diarrhea were found to have cryptosporidiosis (27,28). It appears to cluster in some U.S. urban areas more than in others, which might be due to different levels of contamination of the surrounding environment. In contrast to infection in the United States, more than 50% of patients with AIDS in Haiti and parts of Africa are infected with *Cryptosporidium* (29).

Transmission

Water-borne and person-to-person transmission of *Cryptosporidium* are thought to be the major modes by which the parasite is transmitted to humans. Three well-documented outbreaks of cryptosporidiosis in England were due to contamination of public reservoirs (30–32). During the first of these outbreaks, in 1986, the public water supply was clearly implicated. Oocysts were identified within the intestines of brown trout feeding within the reservoir, as well as from source water and grazing cattle on nearby land. The largest water-borne disease outbreak ever in the United States occurred in Carrollton, Georgia in 1987, affecting an estimated 13,000 individuals (33). Cryptosporidial oocysts were identified in the stool of 39% of persons with diarrhea tested during the outbreak. Oocysts were identified in water samples collected from the city treatment plant and from cattle in the surrounding area.

Person-to-person transmission has been well documented in multiple settings, including clustering of infection among family members of an infected individual and nosocomial acquisition of cryptosporidiosis. During an outbreak in Michigan, 71% of families with an infected child had an additional family member develop cryptosporidial enteritis (34,35).

Persons living in farm communities are at a higher risk of infection due to environmental contamination from infected animals. Small outbreaks among veterinary students and laboratory workers are common (36-38). In one case, *Cryptosporidium* may have been spread from an infected animal to a person via inhalation of infected airborne droplets (39).

In both the western United States and Latin America, *Cryptosporidium* is ubiquitous in water supplies according to sampling studies, though the degree of contamination varies enormously (40). The small size of the oocyst (4 μm) and its relative resistance to chlorination make it difficult to eliminate from water supplies. Sand filtration systems may be more effective at removing oocysts than the various disinfectants commonly used by the water industry (40).

Person-to-person transmission has been documented among both homosexual and heterosexual partners (41,42). It is not unusual for homosexual men who are participating in unprotected anal intercourse to develop multiple concurrent intestinal protozoal infections, including giardiasis, amebiasis, blastocystosis, and cryptosporidiosis.

Nosocomial spread of *Cryptosporidium* to immunodeficient persons may have serious consequences because of the high morbidity and mortality in this group. In one outbreak, six patients in bone marrow transplant unit developed cryptosporidiosis after one of the patients shared a room with another patient with cryptosporidial enteritis before being transferred to the unit (43). In that outbreak, cleaning equipment and washing rags were found to be heavily contaminated by oocysts. The potential of nosocomial spread of *Cryptosporidium* makes it imperative that enteric precautions are observed when infected patients are hospitalized.

Clinical Presentation

Typically, acute infection with *Cryptosporidium* is characterized by watery diarrhea, crampy epigastric abdominal pain, weight loss, anorexia, malaise, and flatulence (44). Diarrhea is the most note-

Table 1.1 Clinical manifestations: human cryptosporidiosis: The New York Hospital (6/82–5/90)

Parameter	AIDS	Non-AIDS
Total no. patients	119	31
Watery diarrhea (BMs/day)	5-20	3-8
Weight loss (kg)	6-23	2-10
Nausea, vomiting	††	†
Abdominal pain	†††	††
Duration of diarrhea (days)	4-720	10-38
Fecal parasite shedding (days)	4-720	10-50
	No. of patients	
Biliary tract involvement	15	0
Pulmonary involvement	2	0
Other intestinal parasites	29	3

worthy symptom and can range from a few loose bowel movements to more than 70 stools (>20 liters) per day (18,45,46) (Table 1.1). Nausea and vomiting may be present. Diarrhea and abdominal pain are usually exacerbated by eating. The spectrum of symptoms from infection is wide; in fact, there are isolated reports of asymptomatic carriage of *Cryptosporidium*. In one study, 12% of patients with vague epigastric abdominal pain or dyspepsia but no diarrhea, who underwent upper endoscopy or endoscopic retrograde cholangio-pancreatography (ERCP), were found to have *Cryptosporidium* oocysts in their duodenal aspirates (47).

The initial physical examination is usually unrevealing (29,48). The patient may have orthostatic hypotension and other signs of dehydration. Low grade fever and mild leukocytosis are not uncommon. The abdomen is usually soft, with mild tenderness on palpation. Laboratory examination often reveals electrolyte disturbances consistent with diarrhea and dehydration. Fecal examination may reveal mucus, but blood and leukocytes are rarely seen. Charcot-Leyden crystals are characteristic of *Isospora belli* infection and amebiasis, but not cryptosporidiosis. Lactose intolerance and fat malabsorption have been well documented, and the D-xylose test is usually abnormal (18,46).

The incubation period may vary from a few days to 2 weeks (49). In immunocompetent individuals the illness is always self-limited, but the diarrhea may be so severe intravenous rehydration is necessary. The duration of illness in immunocompetent hosts ranges from 2 days to a month, but most individuals become asymptomatic within 2 weeks.

Oocyst shedding may vary intermittently from heavy to light during the course of an illness and may persist after clinical resolution. Immunologically healthy individuals all eventually stop excreting oocysts. During the acute illness in healthy persons, more than 80% have detectable oocysts for at least 2 weeks (50). Once symptoms have resolved, it is common for oocysts to be shed for up to a week. In one study, 38% of patients excreted oocysts from 1 day to 15 days (mean 7 days) after symptoms resolved (51). Asymptomatic excretion of oocysts has been reported up to 85 days after the acute illness. Asymptomatic shedding of oocysts in patients who may not even be aware that they had infectious diarrhea is probably important in disease transmission.

Cryptosporidiosis and AIDS

In patients with AIDS, symptoms often begin insidiously with only mild diarrhea, but they increase in severity over time. Most of these patients experience profound, voluminous watery diarrhea ranging from 1 to 25 liters per day, loss of more than 10% of total body weight, and severe abdominal pain (14,29). Without adequate rehydration (either oral or intravenous), the disease may be quickly fatal. Severe malabsorption is the rule, and many patients avoid eating as it worsens the diarrhea and abdominal pain. Occasionally symptoms remit, but these respites are usually brief. Most patients with AIDS never clear the infection and die with cryptosporidial diarrhea.

A subset of patients with AIDS and cryptosporidiosis develop biliary tract involvement (52,53). This complication is invariably associated with severe right upper quadrant pain, nausea, and vomiting. Physical examination reveals right upper quadrant tenderness. Laboratory examination is significant for elevated serum alkaline phosphatase and γ-glutamyltransferase levels. Serum transaminase levels may be mildly elevated, but serum bilirubin levels are usually normal. ERCP reveals dilation of bile ducts with multiple luminal irregularities and distal duct strictures consistent with either partial obstruction or sclerosing cholangitis. Thickening of the gallbladder wall is common. AIDS patients with biliary tract disease may have concomitant infection with cytomegalovirus (CMV) and cryptosporidiosis.

Fulminant cryptosporidiosis is rare among immunodeficient patients without HIV infection, but it does occur. The initial reports of human cryptosporidiosis in immunodeficient patients involved two children with congenital hypogammaglobulinemia and two with impaired humoral and cellular immunity (49). Cryptosporidiosis has also been reported in renal transplant and bone marrow transplant patients (54,55). Patients with reversible immunodeficiencies usually recover when the cause of the immunosuppression is removed.

Cryptosporidial enteritis may occur in HIV-infected individuals who are immunologically competent, as determined by a normal CD4 count. Those individuals have a clinical course identical to HIV-seronegative persons; they are able to clear the infection over a few days to a few weeks. Furthermore, a small number of patients who were followed for 12 months did not relapse when their CD4 counts dropped, suggesting that infection is not always followed by chronic carriage. In a series of 44 patients, it was found that CD4 counts of more than 200 is correlated with the ability to clear cryptosporidial enteritis, whereas individuals who developed chronic unremitting disease had a mean CD4 count of 59 (range 4–150) (56). It appears that the CD4 count may be an accurate prediction of host immunocompetence and the ability to eradicate cryptosporidiosis.

Diagnosis

Diagnosis of cryptosporidiosis is based on identification of the oocyst in stool specimens. At least 15 staining techniques have been tried, but the modified acid-fast stain is optimal because it is both easy to perform and inexpensive (57,58) (Figure 1.4). The acid-fast oocyst stains red with varying intensity and may appear round or crescentic. It is 4 to 6 μm in diameter. Yeasts, which are morphologically similar to *Cryptosporidium* oocysts, are not acid-fast and hence do not stain red (59). A fluorescein-labeled IgG monoclonal antibody (Meridien) that is commercially available and presently being evaluated clinical-

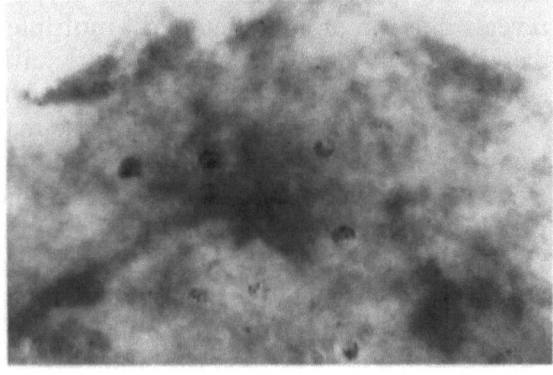

Figure 1.4. *Cryptosporidium* oocysts as they appear on an acid-fast stain of a fecal smear viewed with brightfield microscopy. Oocysts stain red against a blue or green backround, depending on the counterstain. Yeasts, which are approximately the same size as the oocysts, do not stain red.

ly appears to be more sensitive, but not necessarily more specific than the acid-fast stain.

The sensitivity of currently employed diagnostic methods is not known, nor do we know how many stool specimens are needed to confirm that a patient does not have cryptosporidial enteritis. With acute cryptosporidiosis, oocysts are easily detected without concentration techniques. Concentration of oocysts from stool is carried out by flotation in Sheather's sugar solution, hypertonic sodium chloride, zinc sulfate, or Percoll. Concentration techniques are most helpful when examining stool with low numbers of organisms, as in asymptomatic individuals with formed stool. When cryptosporidiosis is a diagnostic consideration, it should be specifically requested as many laboratories do not routinely look for the organism when performing an ova and parasite examination (29). Care should be taken when handling infected specimens in order to prevent leakage or aerosolization, as infection of laboratory workers has been reported (38).

Biopsy of the small intestine is not usually necessary to make the diagnosis of cryptosporidiosis. In fact, the biopsy may be falsely negative owing to the sampling of uninfected areas that result from patchy parasite distribution.

Anti-cryptosporidial IgM, IgG, and IgA can be detected in infected persons by enzyme-linked immunosorbent assay (ELISA) or the indirect fluorescent antibody immunofluorescence assay (IFA). Although useful for epidemiologic evaluation, serologic studies have no role in the diagnosis of acute illness (60).

Treatment

There is currently no effective therapy for cryptosporidial enteritis, even though more than 50 agents have been tried in uncontrolled trials (49). Efforts to develop an in vitro model of infection have been successful, and hopefully these models will be useful for screening and developing new agents.

Development of new agents has been directed at the treatment of patients with AIDS and chronic cryptosporidial enteritis. Cryptosporidiosis in the immunocompetent patient may be treated with supportive care only, consisting in rehydration by either the oral or intravenous route and repletion of electrolyte losses. The severe diarrhea of up to 20 liters a day in patients with AIDS often requires intensive support. Aggressive efforts at oral rehydration should be made with Gatorade, bouillon, or oral rehydration solution, which contains glucose, sodium bicarbonate, and potassium. Often intravenous repletion of fluids and electrolytes is essential to correct losses of bicarbonate, potassium, magnesium, and phosphorus. None of the antimotility agents, such as loperamide, narcotics, or somatostatin, has been shown to be consistently effective, though occasionally they provide temporary relief. Diet should be individualized to fit each patient's preferences, but fatty foods and dairy products should be avoided. Intravenous hyperalimentation may diminish the volume of diarrhea by "putting the bowel to rest" and provides additional nutrition, though it is not appropriate for every patient.

Early reports indicated that the macrolide antibiotic spiramycin was successful in palliating cryptosporidial diarrhea (61,62). A double-blind, placebo-controlled trial of oral spiramycin in patients with AIDS and cryptosporidiosis demonstrated that spiramycin administered at an oral dose of 1 g three times a day was no better than placebo, though both agents led to an amelioration of symptoms in more than 20% of patients (63). Intravenous spiramycin is not effective; it has been evaluated in the AIDS Clinical Trials Group (ACTG) through the National Institutes of Health.

Diclazuril sodium is a veterinary agent with good activity against *Eimeria tenella*, a protozoan parasite closely related to *Cryptosporidium*. More than 50% of patients who received an oral dose of 400 mg a day had a partial response, defined as either a 50% decrease in diarrhea or decrease in stool oocyst number (64). The formulation used was poorly absorbed, therefore a related compound, Letrazuril, is undergoing evaluation in a randomized, placebo-controlled dose escalating trial.

Successful treatment of cryptosporidiosis in a small number of patients with AIDS has been reported with hyperimmune cow colostrum (65,66) and bovine transfer factor (67). Neither the mechanism of action nor the active component in these preparations is known, although there is evidence in a murine model of infection that IgA and IgG1 separated from hyperimmune cow colostrum is effective in preventing *Cryptosporidium* infection (68). Work is ongoing to test these therapies and to define the active components in an in vitro model of infection and in well-controlled clinical trials.

In a small number of patients, the administration of AZT (zidovudine) has led to cessation of diarrhea and clearance of oocysts from stool (69,70). AZT does not have any direct anticryptosporidial activity in an in vitro model of infection (71), but it is believed to act by augmenting immune function. Patients with depressed CD4 counts and cryptosporidiosis should be treated with antiretroviral therapy if possible.

Two oral antimicrobial agents are currently under evaluation for treatment of cryptosporidiosis in patients with AIDS, azithromycin and paromomycin. Paromomycin (Humatin, Parke-Davis) is a poorly absorbed, broad spectrum antibiotic with antimicrobial activity similar to neomycin. At present it is indicated for treatment of intestinal amebiasis. In a retrospective chart review of 12 patients with 23 episodes of gastrointestinal cryptosporidiosis, all patients responded to oral therapy with paromomycin in doses ranging from 1500–2000 mg/day for a median of 14 days (72). 16/23 episodes had a complete response to therapy (symptom improvement, diarrhea eradication and weight gain). 7/23 episodes had a partial response (50% decrease in stool frequency). Many patients initially respond to paromomycin with a decrease in diarrhea, but subsequently relapsed. Paromomycin has potent anti-cryptosporidial activity in vitro (73). Both paromomycin and the oral macrolide antibiotic, azithromycin, are being evaluated in prospective controlled trials.

Animal Models

Experimentally induced infection and disease have been studied in birds (including chickens (74) and turkeys (75), rodents [mice (76,77), guinea pigs (78), and rats], and mammals [rabbits (79), cats (80), dogs (81), pigs (82), sheep (83), goats (84), and cattle (85)].

Existing animal models of *Cryptosporidium* infection are imperfect. Newborn calves are susceptible to infection and develop malabsorptive diarrhea for 1–3 weeks. After the initial bout of cryptosporidiosis, calves are resistant to reinfection (86). Neonatal calves are cumbersome and expensive.

Neonatal Balb/C mice and guinea pigs can be infected with *Cryptosporidium* but do not develop diarrhea or weight loss (87,88). A murine model utilizes adult athymic mice that develop chronic wasting and dehydration (89). Neonatal Balb/C mice treated with either anti-CD4 monoclonal antibodies alone or in conjunction with anti-CD8 monoclonal antibody developed stable infection that resolved once the monoclonal antibody treatments were stopped. Both of these models suggest that T-cell-mediated immunity to cryptosporidiosis is critical for recovery from infection.

In Vitro Infection

The lack of a simple in vitro cultivation system has severely hampered research efforts to study biochemical and metabolic requirements of *Cryptosporidium* and develop effective therapy. Complete

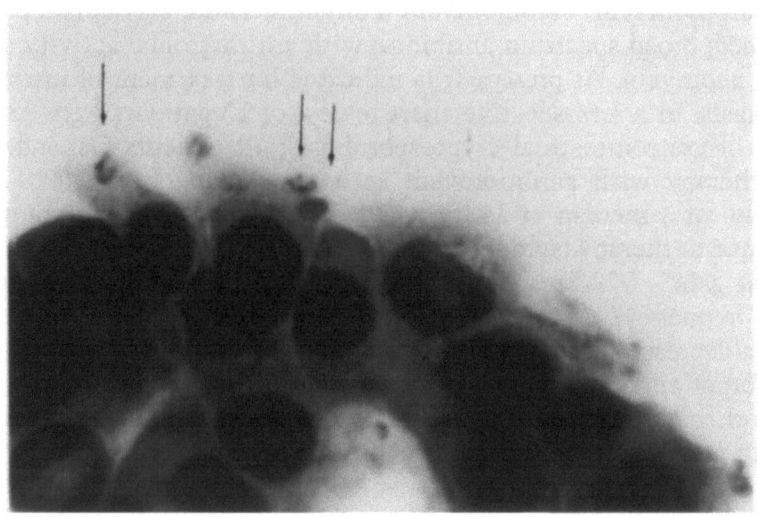

Figure 1.5. Light microscopy of *Cryptosporidium* infection of HT29 cells. Schizonts (arrows) are seen at the periphery of the epithelial cells 24 hours after infection with oocysts. The merozoites within the schizonts stain with hematoxylin. (×1000)

in vitro asexual development of *Cryptosporidium* has been reported in multiple mammalian cell lines, including a human fetal lung cell line, porcine kidney cell line, murine fibroblast cell line, and HT29 and CACO-2 human epithelial cell lines (19,90–92) (Figure 1.5). Although initial in vitro infection is rapid and appears to be similar to in vivo infection, no cultivation system exists, as the level of *Cryptosporidium* replication and production of oocysts in these cell lines is markedly diminished over time. Hopefully, further manipulation of these cell lines will allow the development of high level continuous infection.

To evaluate pharmacologic and immunologic therapies, it is critical to accurately and reproducibly quantify infection. Electron microscopy can definitively identify different life cycle stages, but it is impractical when quantifying infection (Figure 1.6). Parasite stages can be identified with Normanski interference contrast light microscopy, but it requires considerable expertise in parasite morphology. Alternatively, schizonts containing eight distinct merozoites can be easily identified from the host cell and counted by staining with hematoxylin and eosin or Giemsa (19,91). Infection rates in the HT29 cells (HT29.74 subclone) ranged from 50 to 120 schizonts per 1000 cells (19), whereas in a mouse fibroblast cell line infection ranged from 10

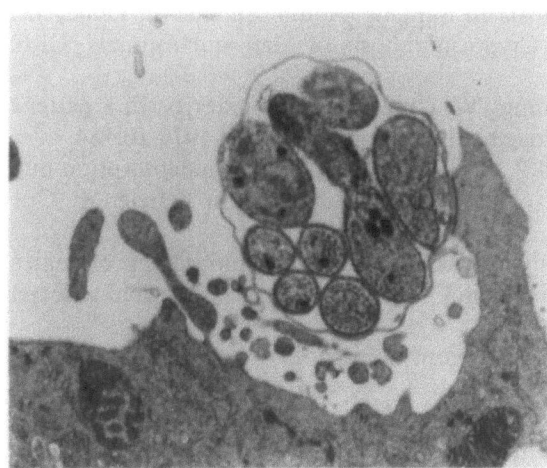

Figure 1.6. Electron microscopy of a *Cryptosporidium* schizont infection of an HT29.74 cell at 24 hours. A dense band separates the parasite from the host cell. Eight merozoites can be identified within the schizont, consistent with a type I schizont. (×15,000)

to 182 schizonts per 50 fields examined under 100× oil immersion objective (91).

An easy to use and efficient in vitro model of *Cryptosporidium* infection allows investigation into parasite invasion of the host cell, intracellular development of the parasite, relevant host defenses, and antimicrobial and immunologic agents that might block invasion and replication.

References

1. Tyzzer EF: A sporozoan found in the peptic glands of the common mouse. Proc Soc Exp Biol Med 1907;5:12.
2. Tzipori S: *Cryptosporidium*: Notes on epidemiology and pathogenesis. Parasitology Today 1985;1:159.
3. Slavin D: *Cryptosporidium* meleagridis (sp. nov.). J Comp Pathol 1955;65:262.
4. Triffit MJ: Observations on two new species of coccidia parasitic in snakes. Protozoology 1925;1:19.
5. Wetzel R: Ein Neues Coccid (*Cryptosporidium* vulpis sp. nov.) aus dem Rotfuchs, Arch Tiereheille 1938;74:39.
6. Panciera RJ, Thomassen RW, Garner FM: Cryptosporidial infection in a calf. Vet Pathol 1971;8:479.
7. Bergeland ME: Necrotic enteritis in nursing piglets. Proc Am Assoc Ve Lab Diagn 1977;20:151.
8. Snyder SP, England JJ, McChesney AE: Cryptosporidiosis in immunodeficient Arabian foals. Vet Pathol 1978;15:12.
9. Nime FA, Burek JD, Page DL, et al: Acute enterocolitis in a human being infected with the protozoan *Cryptosporidium*. Gastroenterology 1976; 70:592.

10. Ungar BLP: Cryptosporidiosis in humans (Homo sapiens). In Dubey JP, Speer CA, Fayer R (eds): Cryptosporidiosis of Man and Animals. CRC Press, Boston, 1990.
11. Lasser KH, Lewin KJ, Ryning FW: Cryptosporidial enteritis in a patient with congenital hypogammaglobulinemia. Hum Pathol 1979;10:234.
12. Sloper KS, Dourmashkin RR, Bird RB, et al: Chronic malabsorption due to cryptosporidiosis in a child with immunoglobulin deficiency. Gut 1982;23:80.
13. Anonymous: Cryptosporidiosis: assessment of chemotherapy of males with acquired immunodeficiency syndrome (AIDS). Morbid Mortal 1982;31:589.
14. Soave R, Armstrong D: *Cryptosporidium* and Cryptosporidiosis. Rev Infect Dis 1986;8:1012.
15. Tzipori S: Cryptosporidiosis in perspective. Adv Parasitol 1988;27:63.
16. Current WL: *Cryptosporidium*: its biology and potential for environmental transmission. CRC Crit Rev Environ Control 1986;17:21.
17. Marcial MA, Madara JL: *Cryptosporidium*: cellular localization, structural analysis of absorptive cell-parasite membrane-membrane interactions in guinea pigs, and suggestion of protozoan transport by M cells. Gastroenterology 1986;90:583.
18. Kotler DP, Francisco A, Clayton F, et al: Small intestinal injury and parasitic diseases in AIDS. Ann Intern Med 1990;113:444.
19. Flanigan TP, Toshika AJI, Marshall R, et al: Asexual development of *Cryptosporidium parvum* within a differentiated human enterocyte cell line. Infect Immun 1991;59:234.
20. Nanduri J, Aji T, Marshall R, et al: Identification and characterization of glycocalyx on the surface of *Cryptosporidium parvum* oocysts. Am Soc Trop Med Hyg 1990;252A.
21. Bossen AN, Britt EM: Cryptosporidiosis in immunocompetent patients. N Engl J Med 1985; 313:1019.
22. Holten-Anderson W, Gerstoft J, Henriksen SA, et al: Prevalence of *Cryptosporidium* among patients with acute enteric infection. J Infect 1984;9:277.
23. Soave R, Ma P: Cryptosporidiosis: traveler's diarrhea in two families. Arch Intern Med 1985;145:70.
24. Ungar BLP, Gilman RH, Lanata CF, et al: Seroepidemiology of *Cryptosporidium* infection two Latin American populations. J Infect Dis 1988;157:551.
25. Mata L, Bolaños, Pizarro H, et al: Cryptosporidiosis in children from some highland Costa Rican rural and urban areas. Am J Trop Med Hyg 1984;33:24.
26. Mata L: *Cryptosporidium* and other protozoa in diarrheal disease in less developed countries. Pediatr Infect Dis 1986;5:117.
27. Smith PD, Lane HC, Gill VJ, et al: Intestinal infections in patients with the acquired immunodeficiency syndrome (AIDS). Ann Intern Med 1988;108:328.
28. Laughon BE, Druckman DA, Vernon A, et al: Prevalence of enteric

pathogens in homosexual men with and without acquired immunodeficiency syndrome. Gastroenterology 1988;94:984.
29. Soave R: Cryptosporidiosis and isosporiasis in patients with AIDS. Infect Dis Clin North Am 1988;2:485.
30. Rush BA, Chapman PA, Ineson RW: *Cryptosporidium* and drinking water. Lancet 1987;2:632.
31. Anonymous: Troubled waters. Lancet 1989;2:251.
32. Smith HV, Girwood RWA, Patterson WJ, et al: Waterborne outbreak of cryptosporidiosis. Lancet 1988;2:1484.
33. Hayes EB, Matte TD, O'Brien TR, et al: Large community outbreak of cryptosporidiosis due to contamination of a filtered public water supply. N Engl J Med 1989;320:1372.
34. Anomymous: Cryptosporidiosis among children attending day-care centers—Georgia, Pennsylvania, Michigan, California, New Mexico. MMWR 1984;33:599.
35. Combee CL, Collinge ML, Britt EM: Cryptosporidiosis in a hospital-associated day care center. Pediatr Infect Dis 1986;5:528.
36. Current WL, Reese NC, Ernst JV, et al: Human cryptosporidiosis in immunocompetent and immunodeficient persons: studies of an outbreak and experimental transmission. N Engl J L Med 1983;308:1252.
37. Anderson BC, Donndelinger T Wilkins RM, et al: Cryptosporidiosis in a veterinary student. J Am Vet Med Assoc 1982;180:408.
38. Blagburn BL, Current WL: Accidental infection of a researcher with human *Cryptosporidium*. J Infect Dis 1983;148:772.
39. Højlyng N, Holten-Andersen W, Jepsen S: Cryptosporidiosis: a case of airborne transmission. Lancet 1987;2:271.
40. Sterling CR: Waterborne cryptosporidiosis. In Dubey JP, Speer CA, Fayer R(eds): Cryptosporidiosis of Man and Animals, 50–57, CRC Press, Boston, 1990.
41. Laughon BE, Druckman DA, Vernon A, et al: Prevalence of enteric pathogens in homosexual men with and without acquired immunodeficiency syndrome. Gastroenterology 1988;94:984.
42. Smith PD, Lane HC, Gill VJ, et al: Intestinal infections in patients with the acquired immunodeficiency syndrome (AIDS): etiology and response to therapy. Ann Intern Med 1988;108:328.
43. Martino P, Gentile G, Caprioli A, et al: Hospital-acquired cryptosporidiosis in a bone marrow transplantation unit. J Infect Dis 1988;158:647.
44. Wolfson JS, Richter JM, Waldron MA, et al: Cryptosporidiosis in immunocompetent patients. N Engl J Med 1985;312:1278.
45. Gillin JS, Shike M, Alcock N, et al: Malabsorption and mucosal abnormalities of the small intestine in the acquired immunodeficiency syndrome. Ann Intern Med 1985;102:619.
46. Kolter DP, Gaetz HP, Lange M, et al: Enteropathy associated with the acquired immunodeficiency syndrome. Ann Intern Med 1984;101:421.
47. Roberts WG, Green PHR, Ma J, et al: Prevalence of cryptosporidiosis in patients undergoing endoscopy: evidence for an asymptomatic carrier state. Am J Med 1989;87:537.

48. Pitlik SD, Fainstein V, Garza D, et al: Human cryptosporidiosis: spectrum of disease: report of six cases and review of the literature. Arch Intern Med 1983;143:2269.
49. Fayer R, Ungar LP: *Cryptosporidium* spp. and Cryptosporidiosis. Microbiol Rev 1986;458.
50. Baxby D, Hart CA, Blundell N: Shedding of oocysts by immunocompetent individuals with cryptosporidiosis. J Hyg (Camb) 1985;95:703.
51. Jokipii L, Jokipii AMM: Timing of symptoms and oocyst excretion in human cryptosporidiosis. N Engl J Med 1986;314:1643.
52. Blumberg RS, Kelsey P, Perrone T, et al: Cytomegalovirus- and *Cryptosporidium*-associated acalculous gangrenous cholecystitis. Am J Med 1984;76:118.
53. Margulis SJ, Honig CL, Soave R, et al: Biliary tract obstruction in the acquired immunodeficiency syndrome. Ann Intern Med 1986;105:207.
54. Collier AC, Miller RA, Meyers JD: Cryptosporidiosis after marrow transplantation: person-to-person transmission and treatment with Spiramycin. Ann Intern Med 1984;101:205.
55. Weisburger WR, Hutcheon DF, Yardley JH, et al: Cryptosporidiosis in an immunosuppressed renal-transplant recipient with IgA deficiency. Am J Clin Pathol 1979;72:473.
56. Flanigan TP, Whalen C, Toerwer J, et al Cryptosporidiosis and CD4 counts, Annals of Int Med 1992;116:840–842.
57. Garcia LS, Bruckner DA, Brewer TC, et al: Techniques for the recovery and identification of *Cryptosporidium* oocysts from stool specimens. J Clin Microbiol 1983;18:185.
58. Ma P, Soave R: Three-step stool examination for cryptosporidiosis in 10 homosexual men with protracted watery diarrhea. J Infect Dis 1983;147:824.
59. Sterling CR, Arrowood MD: Detection of *Cryptosporidium* sp. infection using a direct immunofluorescence assay. Pediatr Infect Dis 1986;5(Suppl):S139.
60. Unger BL, Soave R, Fayer R, et al: Enzyme immunoassay detection of immunoglobulin M and G antibodies to *Cryptosporidium* in immunocompetent and immunocompromised patients. J Infect Dis 1986;153:570.
61. Centers for Disease Control: Cryptosporidiosis: assessment of chemotherapy of males with acquired immunodeficiency syndrome (AIDS). MMWR 1982;31:589.
62. Portnoy D, Whiteside ME, Buckley Ell, et al: Treatment of intestinal cryptosporidiosis with spiramycin. Ann Intern Med 1984;101:202.
63. Soave R: Treatment strategies for cryptosporidiosis. Ann NY Acad Sci (in press).
64. Soave R, Dieterich D, Kotler D, et al: Oral diclazuril therapy for cryptosporidiosis. Sixth Int Conf on AIDS 1990;Th.B.520.
65. Tzipori S, Roberton D, Chapman C: Remission of cryptosporidiosis in an immunodeficient child with hyperimmune bovine colostrum. Br Med J 1986;293:1276.
66. Ungar BLP, Ward DJ, Fayer R, et al: Cessation of *Cryptosporidium*-associated diarrhea in an acquired deficiency syndrome patient after

treatment with hyperimmune bovine colostrum. Gastroenterology 1990;98:468.
67. McMeeking A, Norkowsky W, Klesius PH, et al: A controlled trial of bovine dialyzable leukocyte extract for cryptosporidiosis in patients with AIDS. J Infect Dis 1990;161:108.
68. Fayer R, Guidry A, Blagburn BL: Immunotherapeutic efficacy of bovine colostral immunoglobulins from a hyperimmunized cow against cryptosporidiosis in neonatal mice. Infect Immun 1990;58:2962.
69. Greenberg RE, Mir R, Bank S: Resolution of intestinal cryptosporidiosis after treatment of AIDS with AZT. Gastroenterology 1989;97:1327.
70. Connolly, GM, Dryden MS, Shanson, et al: Cryptosporidial diarrhea in AIDS and its treatment. Gut 1988;29:593.
71. Flanigan TP, Marshall R, Redman D, et al. In Vitro Screening of Therapeutic agents against *Crytosporidium*: Hyperimmune Law Colostrum is Highly Inhibitory. J. Protozology 1991:39;6:255–2275.
72. Gathe J, Piot D, Hawkins K, et al. Treatment of gastrointestinal cryptosporidiosis with paromomycin. Sixth Annual conference on AIDS, 1990:2121.
73. Marshall MS and Flanigan TP. Paromomycin inhibits cryptosporidiosis infection of a human enterocyte cell line. JID, 165:772–774, 1992.
74. Itakura C, Goryo M, Umemura T: Cryptosporidial infection in chickens. Avian Pathol 1985;13:487.
75. Hoerr FJ, Ranck FM, Hastings TF: Respiratory cryptosporidiosis in turkeys. J Am Vet Med Assoc 1978;173:1591.
76. Heine J, Moon HW, Woodmansee DB: Persistent Cryptosporidium infection in congenitally athymic (nude) mice. Infect Immun 1984;43:856.
77. Tyzzer EE: *Cryptosporidium* parvum (sp nov.) a coccidium found in the small intestine of the common mouse. Arch Protistenkd 1912;26:394.
78. Argenzio RA, Liacos JA, Levy ML: Villous atrophy, crypt hyperplasia, cellular infiltration, and impaired glucose-Na absorption in enteric cryptosporidiosis of pigs. Gastroenterology 1990;98:1129.
79. Anman LR, Yakeuchi A: Spontaneous cryptosporidiosis in an adult female rabbit. Vet Pathol 16:1989
80. Iseki M: *Cryptosporidium* felis Sp. N. (Protozoa: Eimeriorina) from the domestic cat. Jpn J Parasitol 1979;28:285.
81. Sisk DB, Gosser HS, Styer EL: Intestinal cryptosporidiosis in two pups. J Am Vet Med Assoc 1984;184:835.
82. Moon HW, Bemrick WJ: Fecal transmission of calf cryptosporidia between calves and pigs. Vet Pathol 1981;18:248.
83. Angus KW, Appleyard WT, Menzies JD, et al: An outbreak of diarrhea associated with cryptosporidiosis in naturally reared lambs. Vet Rec 1982;110:129.
84. Tzipori S, Angus KW, Gray EW, et al: Diarrhea in lambs experimentally infected with *Cryptosporidium* isolated from calves. Am J Vet Res 1981;42:1400.
85. Morin M, Lariviere S, Lallier R, et al: Neonatal calf diarrhea: pathology and microbiology of spontaneous cases in dairy herds and incidence of the enteropathogenes implicated as etiological agents. In: Proceedings,

2nd International Symposium on Neonatal Calf Diarrhea, 1978, p. 347.
86. Harp JA, Woodmansee DB, Moon HW: Resistance of calves to *Cryptosporidium parvum*: effects of age and previous exposure. Infect Immun 1990;58:2237.
87. Perryman LE: Cryptosporidiosis in rodents: In Dubey JP, Speer CA, Fayer R (eds): Cryptosporidiosis of Man and Animals, 125–132, CRC, Boca Raton, FL, 1990.
88. Chrisp CE, Reid WC, Rush HG: Cryptosporidiosis in guinea pigs: an animal model. Infect Immun 1990;58:674.
89. Ungar BLP, Burris JA, Quinn CA, et al: New mouse models for chronic Cryptosporidium infection in immunodeficient hosts. Infect Immun 1990;58:961.
90. Current WL, Haynes TB: Complete development of *Cryptosporidium* in cell cultures. Science 1984;224:603.
91. McDonald V, Stables R, Warhurst DC, et al: In vitro cultivation of *Cryptosporidium parvum* and screening for anticryptosporidial drugs. Antimicrob Agents Chemothe 1990;34:1498.
92. Datry A, Danis M, Gentilini M: Development complet de *Cryptosporidium* en culture cellulaire: applications. Med Sci 1989;5:762.

2
Toxoplasmosis: An Update on Clinical and Therapeutic Aspects

Catherine F. Decker and Carmelita U. Tuazon

Once considered a rare central nervous system (CNS) infection in the immunocompromised host, toxoplasmosis has emerged as a frequently encountered infectious disease since the onset of the acquired immunodeficiency syndrome (AIDS) epidemic. A resurgence of attention has focused on the pathogenesis, diagnostic methods, and approaches to treatment of the disease. Prior to 1980 toxoplasmic encephalitis occurred sporadically in the immunocompromised host predominantly in patients with underlying malignancies of the reticuloendothelial system and cardiac transplants (1). In the AIDS population, toxoplasmic encephalitis is the most common cause of an intracerebral mass lesion. It has been proposed that 5-10% of AIDS patients in the United States and 25-40% of AIDS patients in Western Europe will develop CNS disease (2).

In the normal host, acute infection is asymptomatic in older children and adults and is discovered only by the presence of antibodies to *Toxoplasma gondii*. Symptomatic toxoplasmosis can be categorized into four major clinical syndromes: congenital, ocular, lymphadenopathic, and infection in the immunocompromised host. In the immunocompromised patient with malignancy or the transplant patient and the infant in utero, acute infection or reactivation of latent infection may cause severe life-threatening disease. In the AIDS patient the infection is a result of reactivation of latent infection.

The Organism

Toxoplasma gondii, an obligate intracellular parasite is worldwide in distribution and can infect all mammalian species. The life cycle of *T. gondii* and its mode of transmission has been well described. The parasite exists in three forms: tachyzoite (proliferative), tissue cyst, and oocyst, all of which are potentially infectious. Tachyzoites can infect cells or form a membrane within the cell to produce a cyst. The cyst may vary in size and contain hundreds of organisms that persist for the life of the host. These cysts are resistant to freezing, but they are destroyed by heating above 60°C and desiccation. Felines are the definitive host where the sexual phase of the disease occurs. After becoming infected, the cat can shed sporocysts in feces for 1–3 weeks. Depending on the temperature, these cysts can sporulate in the soil and can be infectious for up to 1 year.

Toxoplasma infection is generally transmitted to humans by ingestion of tissue cysts in raw or undercooked meat. Trophozoites are released through digestion of the cyst by peptic enzymes. Organisms invade the mucosa of the gastrointestinal tract, replicate and disrupt host cells, and disseminate throughout the body. Proliferation of tachyzoites often produces foci of necrosis surrounded by a mononuclear reaction. Intact humoral and cellular immunity usually halts tissue destruction. Tissue cysts can be demonstrated as early as 1 week after infection and are found most commonly in the brain and cardiac and skeletal muscle. They remain viable for the life of the host and have the potential to reactivate (3).

The increase in incidence of toxoplasmic encephalitis has regenerated interest in methods for in vitro isolation of this organism (4). Traditionally, the organism has been isolated by inoculating tissue specimens into the peritoneal cavity of mice. Tissue specimens should not be treated with formalin, as it is lethal to the organism. The peritoneal fluid is then examined 6–10 days after inoculation. Most strains of *Toxoplasma* isolated from humans are avirulent for mice; and if mice should survive at 6 weeks, the presence of antibodies is determined. If antibodies are present, demonstration of the organism in organs such as brain, liver, or spleen establishes a definitive diagnosis. If cysts are not seen, suspensions from tissue are subinoculated into fresh mice (3).

Inoculation of the buffy coat of a patient's blood into tissue culture is another technique used to isolate *T. gondii* from patients with active infection. The isolation of the organism from blood or body fluid confirms that infection is acute. *Toxoplasma* forms plaques in tissue culture of human foreskin fibroblasts and cultured cell lines. When stained with Wright-Giemsa stain, these plaques demonstrate necrotic and heavily infected cells as well as numerous extracellular tachyzoites. As few as three organisms can cause plaque formation in human fibroblast monolayer as early as 4 days, but it varies with the virulence of the strain of *T. gondii* (4). The sensitivity of tissue culture technique in relation to traditional mouse inoculation has yet to be determined. Isolation of the organism may take 6 days to 6 weeks after tissue culture or mouse inoculation, and waiting for results is not practical in the initial management of the patient (5).

Epidemiology and Transmission

The major mode of transmission of *T. gondii* to humans is the oral route: tissue cysts in meat. The number of tissue cysts in meat is high, particularly in undercooked or raw meat, which is customarily consumed in certain cultures. In accordance with this practice, 70–96% of the adult population in Germany and France have serologic evidence of toxoplasmosis. In the United States 20–70% of adults are infected. In general, the incidence of seropositivity varies from population to population and according to geography. It also appears to increase with age but does not vary between sexes. Because sporocysts can contaminate soil and vegetation, the disease may be acquired by vegetarians. There have been reports of transmission by unpasteurized goat milk (6). There has been no substantial evidence to support human-to-human transmission, other than the congenital route, but outbreaks in families have been noted (7). Accidentally acquired infection has occurred in laboratory workers (8). The congenital route may be responsible for transmission when the mother develops acute infection during pregnancy. This risk appears to correlate with geographic location, as moving from an area of low prevalence of infection to an area of high prevalence before the time of childbearing may increase the risk (9). Transmission has also occurred via blood products (10) and through transplantion of an organ from a seropositive donor to a seronegative recipient (11).

Pathogenesis and Host Defenses

Much controversy exists as to immune mechanisms operative in latent infection as well as in reactivation disease, particularly encephalitis in the immunocompromised host. Both humoral and cellular immunity play major roles. Investigators have focused on the protective role of γ-interferon and the role of CD4 T cell function.

In the murine model, γ-interferon has been shown to be of paramount importance in resistance against toxoplasmic infection (12). Similarly, this macrophage-activating factor has been demonstrated also to be important in the prevention of reactivation of latent chronic infection in the form of toxoplasmic encephalitis. Suzuki et al. demonstrated that mice chronically infected with *T. gondii* after treatment with monoclonal antibody against γ-interferon developed severe toxoplasmic encephalitis (13). Histopathologically, tissue sections from mice treated with γ-interferon monoclonal antibody revealed areas of acute focal inflammation in which immunoperoxidase staining revealed tachyzoites and toxoplasmic antigens surrounding the cysts. Tachyzoites were also present outside the cyst, suggestive of cyst rupture, which may support a possible pathogenesis of toxoplasmic encephalitis in the immunocompromised host. γ-Interferon activates macrophages invivo and invitro to kill *T. gondii* (12,14). Patients with AIDS have impaired production of γ-interferon, which may account for the increased incidence of toxoplasmic encephalitis (15). In a recent report, gamma interferon levels were significantly lower in AIDS patients with toxoplasma encephalitis than in immunocompetent patients with toxoplasma lymphadenopathy or healthy controls. This finding lends additive support to the protective role of this lymphokine in human toxoplasmosis (16). Similarly, studies of mononuclear phagocytes among neonates have demonstrated a defect in production and response to γ-interferon (17).

These are conflicting views on the role of CD4 T-cell function in the reactivation of *T. gondii*. Vollmer et al. (18) demonstrated that mice chronically infected with *T. gondii* and treated with a monoclonal antibody against the cell surface membrane glycoprotein L3/T4 (CD4) of T-lymphocytes developed toxoplasmic encephalitis similar to patients with AIDS. Israelski et al. directly contradicted this finding when they reported that after treatment with the same monoclonal antibody the inflammatory response was decreased in the brains of mice chronically infected with *T. gondii* (19). However, after treatment was discontinued, a recrudescence of the inflammatory process characterized by microabscesses and necrosis was observed. The investigators postulated that the development of toxoplasmic encephalitis might result from alteration in the ratio of lymphocyte subpopulations responsible for toxoplasmic infection persisting in a latent stage. Further investigations are needed.

Other immune responses have been postulated to be important in hindering the proliferation of this parasite and maintaining the latent phase in the immunocompetent host. *Toxoplasma* blocks lysosome–phagosome fusion and prevents acidification of the phagosome interfering with macrophage killing (20). However, these parasites are susceptible to both oxygen-dependent and oxygen-independent host cell microbicidal mechanisms (21). Ferguson et al. demonstrated in elegant immunocytochromal and ultrastructural studies the sequelae of tissue cyst rupture in the brains of asymptomatic chronically infected mice (22). In immunocompetent mice chronically infected with *T. gondii*, cyst rupture occurred intermittently but rarely. Characteristic of a ruptured cyst was a rapid cell-mediated immune response resulting in microglial or inflammatory nodules containing *Toxoplasma* antigens. Macrophages were observed to phagocytize *Toxoplasma*, and parasite multiplication or new cyst formation was not noted. It may be postulated that in immunocompromised pateints such as patients with AIDS or patients receiving immunosuppression, unchecked parasite proliferation after cyst rupture could result in the development of large necrotic intracerebral lesions.

In addition to the immunologic status of the patient, the strain of *T. gondii* that causes infection may be a factor in the morbidity and severity of the disease. This factor has been demonstrated in the murine model (23).

By a yet undefined mechanism, pregnancy appears to increase susceptibility to toxoplasmosis, and a marked increase in mortality has been noted in the murine model (24). An increase in reports of clinical toxoplasmosis has also been reported in transplant recipients. Although some of these infections may have been acquired from the donated organ, most are the result of reactivation of latent *T. gondii* due to the use of immunosuppressive drugs (25).

Clinical Presentation

Although toxoplasmosis is generally subclinical in nature, in the immunocompetent host there are three major settings where the acquisition of *Toxoplasma gondii* may cause disease: (1) in utero; (2) ocular involvement; and (3) acute acquired toxoplasmosis.

Women with chronic or latent toxoplasmosis acquired prior to pregnancy do not transmit the infection to their fetuses, nor does toxoplasmosis produce recurrent abortion in women with persistently positive antibody titer. If toxoplasmosis is acquired during pregnancy, transplacental infection of the fetus can occur. Between 0.2% and 0.5% of pregnancies worldwide are complicated by acquired toxoplasmosis (26). About 40% of these pregnancies have fetal infection, and more than half the infants with congenital toxoplasmosis present with subclinical infection. The clinical spectrum of symptomatic congenital infection includes fetal death, neurologic damage including cerebral calcifications, seizures, retardation, hydrocephalus, microcephaly, chorioretinitis, fever, hepatosplenomegaly, and rash. Infections acquired during the second month through the end of the sixth month of pregnancy are more likely to cause severe disease than infections acquired during the third trimester (27). Congenital toxoplasmosis can be diagnosed by serology or by identification of the organism in the placenta or fetal tissues. Infants who survived may benefit from treatment with pyrimethamine and sulfonamides. Reports suggest that most infants with subclinical infection at birth eventually develop signs and symptoms of congenital toxoplasmosis (28).

Ocular toxoplasmosis is considered to be one of the most common causes of posterior granulomatous uveitis. Although ocular inflammation is rare in acquired toxoplasmosis, chorioretinitis could result from reactivation of a congenital infection and is most commonly seen in older children and young adults. Typically, lesions exhibit yellow-white fluffy exudates clustered in the posterior pole. Visual impairment is the most common symptom, but pain and photophobia are frequent manifestations. Ocular toxoplasmosis is a clincal diagnosis that depends on the morphology of the lesion in the presence of a positive toxoplasmal serology. Other conditions included in the differential diagnoses are tuberculosis, syphilis, histoplasmosis, candidiasis and leprosy (29). Relapses of chorioretinitis are frequent despite therapy.

Lymphadenopathy is the most common symptom of acute toxoplasmosis. It may be localized or generalized, usually involving the cervical lymph nodes, particularly in the posterior cervical region. Lymphadenopathy may persist for months and may be the only clinical manifestation of disease, although fever, malaise, myalgias, and pharyngitis may also be present. A maculopapular rash and occasionally hepatosplenomegaly are present in a few patients. The acute illness often mimics infectious mononucleosis, cytomegalovirus, lymphoma, or less often sarcoidosis or cat scratch disease. Serology is important in the diagnosis of this illness, and lymph node biopsy may be highly suggestive (30). Symptoms may persist for weeks to months, although the illness is usually self-limited and does not require therapy. A chronic form of lymphadenopathy may persist (31). Rarely, complications of acquired toxoplasmosis occur, including

Table 2.1 Clinical presentation of patients with AIDS and CNS toxoplasmic encephalitis

Parameter	Carrazana et al. (34) (no.)	Leport et al. (55) (no.)	Levy & Bredesen (33) (no.)	Wanke et al. (32) (no.)	Haverkos (35) (no.)	Navia et al. (36) (no.)	Wong et al. (52) (no.)
No. pts. evaluated	26	35	35	14	61	27	7
Seizures	8 (31)[a]	14 (40)	13 (24)	6 (43)	22 (35)	4 (15)	1 (14)
Focal neurologic signs	19 (73)	17 (49)	53 (100)	4 (30)	36 (58)	24 (89)	5 (71)
Mental status change	13 (50)		37 (10)	2 (14)		17 (63)	2 (29)
Coma		8 (23)				8 (30)	
Confusion		11 (31)			27 (37)		
Meningeal symptoms	4 (11)				2 (4)		2 (29)
Psychosis				1 (7)	1 (2)		
Fever	1 (4)	26 (74)	1 (2)	8 (57)	6 (19)	15 (56)	
Headache	19 (73)		24 (45)		27 (44)	15 (56)	

[a] Numbers in parentheses are percents.

unilateral chorioretinitis, pericarditis, myocarditis, pneumonitis, myositis, or meningoencephalitis. Family members of a patient with acute toxoplasmosis are at risk for acquiring the infection, presumably from a common source (7).

In the immunodeficient patient, in particular the patient with AIDS, early diagnosis of toxoplasmosis is essential to reducing morbidity and mortality. Patients may present with focal or generalized neurologic abnormalities. Focal abnormalities include focal seizures, hemiparesis, hemisensory loss, cranial nerve palsies, diplopia, cognitive impairment, personality changes, and headaches (1). A review of the literature reporting on the clinical presentation of patients with AIDS and toxoplasmic encephalitis (Table 2.1) noted that focal and neurologic abnormalities in 30–100% and seizure activity was observed in 25–40% of patients (32–36). The incidence of fever has varied widely in reports from 2% to 74%, but several studies admit to inadequate evaluation of this symptom. The most commonly noted symptoms included headache, disorientation, and hemiparesis (35). Patients may also display ataxia, aphasia, visual field loss, cranial nerve deficits, or movement disorders (36). There have been reports of toxoplasmic chorioretinitis preceding or accompanying CNS disease (37). Involvement of the heart (38), lungs (39), and testes (40) has also been reported in patients with AIDS plus disseminated toxoplasmosis. The syndrome of inappropriate antidiuretic hormone secretion may be the primary manifestation or late complication of CNS toxoplasmosis (41).

Diagnosis

The initial diagnosis of toxoplasmic encephalitis and the decision to initiate therapy are based on a high index of suspicion and findings on radiographic imaging of the brain. Computed tomography (CT) and magnetic resonance imaging (MRI) have been valuable for diagnosing and assessing response to therapy in patients with toxoplasmic encephalitis. The density and contrast enhancement of cerebral toxoplasmosis lesions have demonstrated substantial variability. Contrast enhancement of the lesions has been reported in 80–94% of cases. Typically, the CT scan demonstrates multiple enhancing mass lesions (Fig. 2.1). MRI of the head appears to be more sensitive than the CT scan and may demonstrate lesions not detected on CT scan. In a review, correlation of cranial MRI with CT was performed in 32 patients. MRI was shown to be superior for detecting lesions, demonstrating more lesions than CT in 14 studies (44%) and equivalent information in 18 studies (56%). In no case did CT demonstrate

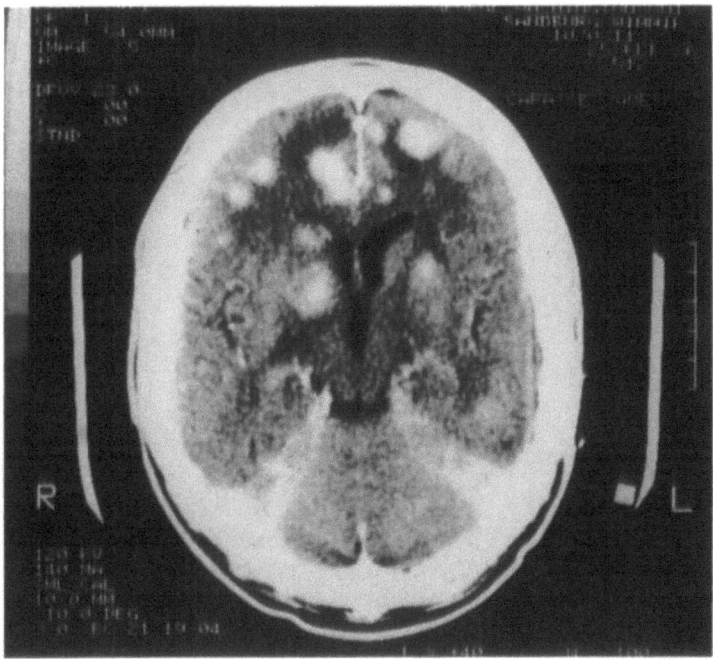

Figure 2.1. CT scan of a patient with AIDS demonstrating multiple lesions in the brain.

lesions not detected in MRI (42). MRI invariably demonstrates multiple mass lesions as in Figure 2.2. Most of the lesions are noted in cerebral hemispheres with a predilection for the basal ganglia and hemispheric corticomedullary junction (43). Lesions have been reported as solitary or multiple and isodense or hypodense. Mild to severe edema is present on the CT scans of almost all patients.

In a comparison of histopathologic findings with CT and MRI studies in patients with AIDS presenting with neurologic symptoms, MRI reflected the extent and distribution of CNS lesions more accurately. In six patients with toxoplasmic encephalitis, MRI consistently demonstrated the presence of multiple bilateral lesions that correlated with autopsy findings. In comparison, CT scan demonstrated the presence of bilateral lesions in two of these patients. In the other four patients, CT scans were normal in one, demonstrated a single lesion in two, and revealed two lesions in one (44).

Findings on CT scan and MRI are not pathognomonic for toxoplasmic encephalitis, as they are for other disease entities; for example, lymphoma, fungal abscess, tuberculoma, and Kaposi's sarcoma may present similarly, especially as a single lesion on MRI scan.

Figure 2.2. MRI showing multiple mass lesions.

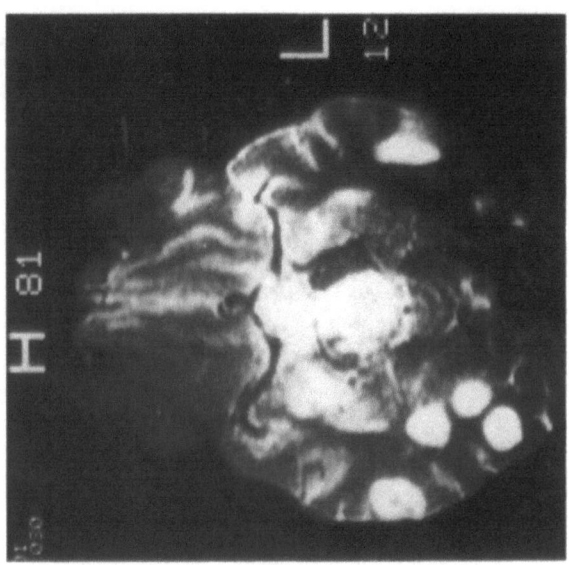

A review of CT scans of 200 patients with AIDS revealed focal lesions in 26% of cases. Of these patients with focal abnormalities on CT scan, 50–70% were reported to have toxoplasmic encephalitis, 10–25% primary CNS lymphoma, and 10–22% progressive multifocal leukoencephalopathy; 10% had either nondiagnostic biopsies or other diseases (45).

Serologic testing rarely confirms the diagnosis of toxoplasmic encephalitis in an AIDS patient (46). Although immunoglobulin G (IgG) antibodies are predictably positive, this finding has been reported in 40–50% of healthy individuals in the United States with chronic inactive infections (3). In France, approximately 80% of the general population has evidence of previous infection. In one study, IgM antibody to *T. gondii* has been demonstrated in only 20% of patients with AIDS and toxoplasmic encephalitis. Patients who are recipients of organ transplants and are seropositive for antibody to *T. gondii* may have a rise in serologic titers with the presence of IgG or IgM antibodies. This rise in antibodies may not necessarily reflect active *Toxoplasma* infection.

Increased antibodies in the cerebrospinal fluid (CSF) may be helpful in establishing the diagnosis of toxoplasmic encephalitis (47). Local antibody production may be dependent on the proximity of the encephalitic process to the meninges. Antigen detection has not been proved to be useful in the diagnosis of toxoplasmic encephalitis. Western blot analysis to determine antibodies recognizing specific

Toxoplasma antigens in AIDS patients with toxoplasmic encephalitis demonstrated a wide variation of antibody response that may represent antigen diversity of Toxoplasma strains (48). A new diagnostic method utilizing polymerase chain reaction to detect T. gondii may provide additional sensitivity to the diagnosis. It may be helpful for detecting the parasite in CSF or in the buffy coat of peripheral smears (49,50).

In the immunocompetent patient, serologic testing is the mainstay of diagnosis. Toxoplasmic lymphadenitis is an almost exclusive manifestation of acute acquired Toxoplasma infection. A negative result in the dye test or IgG antibody excludes the diagnosis except in rare settings where humoral response may be delayed. The DS-IgM ELISA is more sensitive and specific than the IgM-IFA test (51). A negative IgM IFA does not exclude acute Toxoplasma infection, whereas a negative DS-IgM ELISA markedly reduces the likelihood of acute acquired Toxoplasma infection. Acute acquired infection can be diagnosed by seroconversion from a negative to a positive IgG antibody or by demonstrating a serial two-tube fourfold rise from a low to a high titer in sera tested in parallel and obtained at 3-week intervals.

A definitive diagnosis can be made with demonstration of the organisms from brain tissues obtained by diagnostic biopsy (Fig. 2.3). Use of image-guided stereotaxic brain biopsy appears to be safer and a more effective means for diagnosis. Direct examination for parasites of Wright-Giemsa-stained smears of brain aspirate or other tissue specimens or of centrifuged CSF sediment may be diagnostic (52). The sensitivity of this technique is low, however, and additional specimens derived from mouse inoculation and tissue culture techniques may be helpful.

Because of technical difficulties arising from the anatomic location of the mass lesion, surgical morbidity, and sampling error, it has been recommended and is accepted general practice that patients who have AIDS with enhancing CNS mass lesions are presumed to have toxoplasmosis and so are initiated on empiric therapy. The patient is carefully observed, and if there is no radiographic or clinical improvement after 7–10 days of therapy, a brain biopsy should be performed to rule out an alternative diagnosis. Demonstration of T. gondii confirms the diagnosis (Fig. 2.4). Toxoplasmic encephalitis has been reported to occur concurrently with other pathogens, e.g., *Cryptococcus neoformans, Mycobacterium tuberculosis,* and *Aspergillus* sp. It is equally important to recognize that these radiographic findings on CT scan or MRI studies are not pathognomonic for toxoplasmosis and may be observed with other disease processes, e.g., lymphoma, Kaposi's sarcoma, mycobacteriosis, cryptococcosis, candidiasis, and aspergillosis.

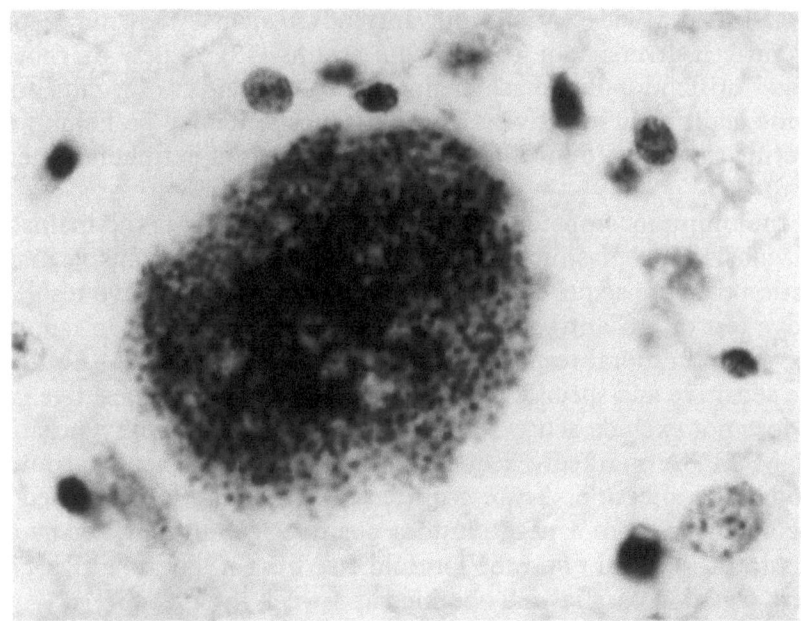

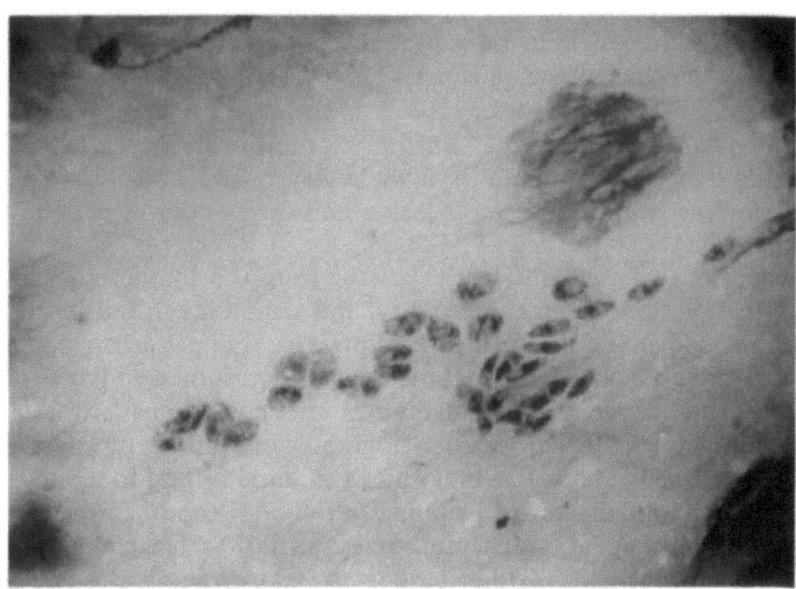

Figure 2.3. Brain aspirate obtained by stereotaxic biopsy demonstrating *T. gondii*. (A, ×1000; B, ×40)

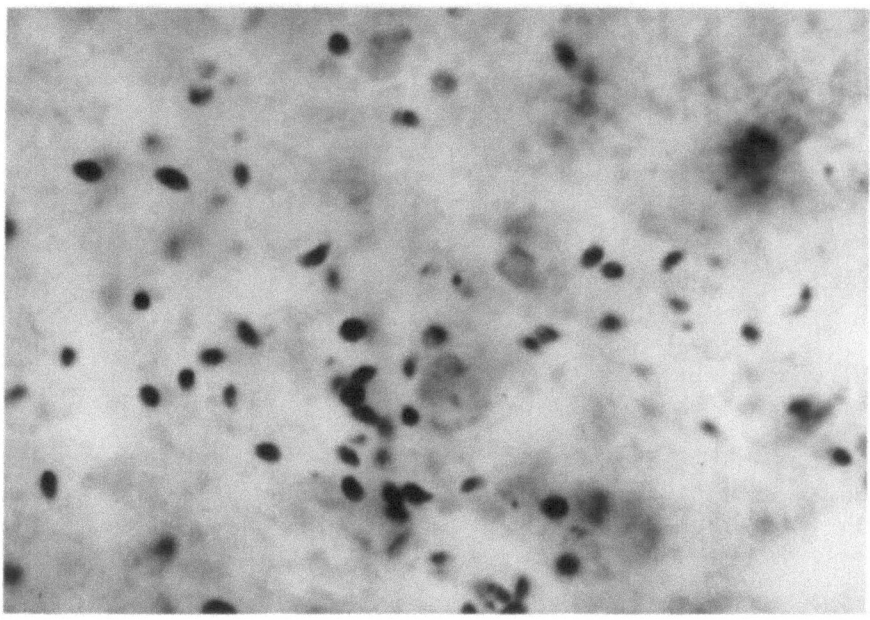

Figure 2.4. Histopathologic section demonstrating *T. gondii* in brain tissue obtained by open biopsy. (×40)

Histopathologically, lesions caused by *T. gondii* produce necrotizing granulomas with thin capsules and minimal inflammation. It may be difficult to demonstrate tachyzoites in tissue. One report of toxoplasmic encephalitis in AIDS patients was successful in only one-half of the specimens obtained by open brain biopsy (52). *T. gondii* cysts and tachyzoites may be best demonstrated by immunohistopathologic staining of brain tissue with the peroxidase antiperoxidase stain, which uses antisera to *Toxoplasma* (46,53). This method is more sensitive for making the diagnosis of *T. gondii* encephalitis than is direct visualization.

Treatment

The increased incidence of toxoplasmic encephalitis during the AIDS era has allowed for critical assessment of treatment regimens and the introduction of new therapeutic agents.

Initiation of early and aggressive antimicrobial therapy usually results in a dramatic clinical and radiographic response within 7–10 days. The combination of pyrimethamine and sulfadiazine has been

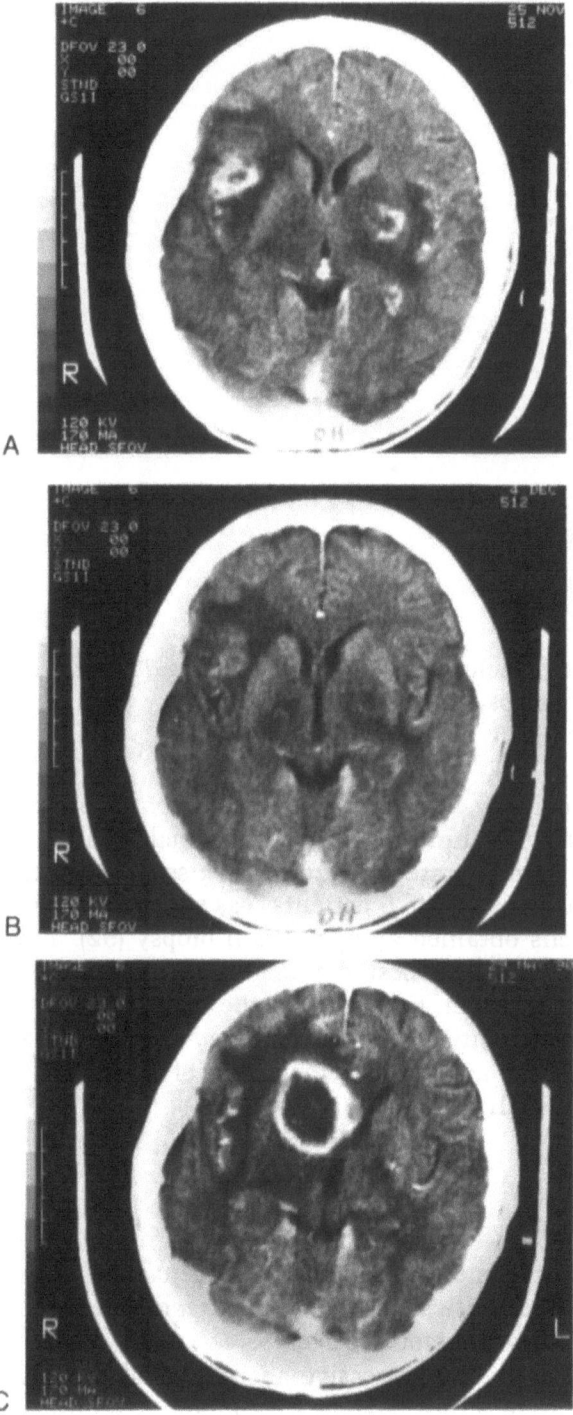

demonstrated to be the most efficacious regimen for the treatment of patients with AIDS and toxoplasmic encephalitis (54,55). The antimicrobial combination is synergistic by sequential blocking of folic acid metabolism of the proliferative stage of *T. gondii*. *T. gondii* lacks the transport system for the uptake of folate that mammalian cells exhibit and must synthesize their folates de novo. Several of the enzymes of this folate pathway include dihydropteroate synthetase (DHPS) and dihydrofolate reductase inhibitor (DHFR), which are the targets for competitive inhibition by antimicrobial agents (2). Sulfadiazine is a DHPS inhibitor, and pyrimethamine is a DHFR inhibitor. The activity of these drugs is limited to the replicating tachyzoite, and the cyst form remains a viable source of organisms that is able to rupture and reinitiate the process. Such a latent process probably accounts for the high incidence of relapses among patients with AIDS.

The major problem in the treatment of toxoplasmic encephalitis, similar to other opportunistic infections in patients with AIDS, is the high relapse rate (Fig. 2.5). Daily maintenance therapy for documented disease should be continued for the patient's lifetime. LePort et al. treated 35 AIDS patients who had clinical and CT scan findings consistent with toxoplasmic encephalitis with combination pyrimethamine and sulfadiazine (55). The patients were followed for a 30-month period. During the first 2 months of therapy, 31 showed improvement. Of 24 patients evaluable for long-term therapy, 14 (58%) achieved complete resolution (55) (Table 2.2) Approximately 40% of patients begun on pyrimethamine and sulfadiazine eventually discontinue treatment because of drug toxicities, including bone marrow suppression, fever, or rash (35).

The most common toxicity observed with pyrimethamine is bone marrow suppression manifesting as neutropenia and thrombocytopenia, which is dose-related. Folinic acid is administered to counteract toxicity from pyrimethamine. Skin rash and fever are also frequently observed as adverse reactions to drugs used for the treatment of toxoplasmic encephalitis. In the patient intolerant to sulfadiazine, clindamycin has been used in combination with pyrimethamine and is currently proposed as second-line therapy in patients with AIDS (56). LePort et al. demonstrated in an uncontrolled study that the combination of clindamycin with pyrimethamine was effective in

---

Figure 2.5. (A) CT scan showing multiple mass lesions in an AIDS patient with toxoplasmic encephalitis. (B) CT scan after 9 days of therapy showing resolution of lesions. (C) CT scan demonstrating recurrence of lesion after discontinuation of therapy.

Table 2.2 Outcome of patients with AIDS treated for toxoplasmic encephalitis with pyrimethamine and sulfadiazine

Study	Duration (days)	Clinical response		
		None	Partial	Complete
LePort et al. (55) (n = 35)	1–70	4	21	10
Wanke et al. (32) (n = 13)	7–180	2	3	8
Wong et al. (52) (n = 71)	25–100	3	1	3
Navia et al. (36) (n = 26[a])	1–330	4	7	10

[a] Five patients received no treatment.

eight of nine patients; but on long-term evaluation a relapse rate of 33% was noted with a follow-up of 8 months (57). Recently, results of a large collaborative randomized trial suggested that the relative efficacy of pyrimethamine and clindamycin approximately equals that of pyrimethamine and sulfadiazine in the treatment of toxoplasma encephalitis in patients with AIDS (58).

The side effects of currently available drugs and the high rate of relapse has led to the search for additional agents with activity against *T. gondii*. Kovacs et al. (59) evaluated the ability of *T. gondii* to incorporate tritiated paraaminobenzoic acid during de novo folate synthesis in the presence of folate inhibitors. In addition to pyrimethamine and sulfonamides, dapsone and trimetrexate are potent folate inhibitors.

Other interesting agents that demonstrate in vivo and in vitro activity against *T. gondii* are the macrolides (58,60–61) and the trioxane compounds (62). Studies have demonstrated that macrolides are able to block the nucleotide synthesis of intracellular parasites (62). The exact mode of action is unclear, but macrolides must penetrate the macrophage to exert their activity against intracelluar *T. gondii*. Of the macrolides tested, roxithromycin reached considerably higher intracellular/extracellular concentration ratios. Recently, there has been scattered reports of the use of macrolides alone or in combination for the treatment of toxoplasma encephalitis in patients with AIDS. In a small pilot study of 13 AIDS patients with toxoplasma encephalitis, the combination of pyrimethamine and clarithromycin was shown to be comparable to the conventional regimens. The clinical and computed tomography scan responses in 8 evaluable patients at six weeks of treatment were 80% and 50% respectively (63). Azithromycin has also been used as a single agent in the treatment of cerebral toxoplasmosis (64).

A new hydroxynapthaquinone compound, 566C80 has also been found to be effective in vitro and in vivo against *T. gondii* and is currently undergoing investigation in clinical trials.

The search for additional agents with activity against *T. gondii* has also led to improvement in the methods used for in vitro screening of drugs. With the cyst form of *T. gondii* considered responsible for recrudescence of latent infection, a method has been developed to evaluate the effect of drugs on the cyst forms. The most active compounds against the cyst forms in vitro were azithromycin, aprinocid-N-oxide and the hydroxynapthoquinone, 566C80 (65).

With the production of recombinant lymphokines that enhance cellular immunity, immunotherapy alone or in combination with specific chemotherapy may be useful for the treatment and prevention of relapse in patients with AIDS (1). As mentioned earlier, γ-interferon afforded protection in mice infected with *T. gondii*. γ-Interferon has also been shown to act synergistically with anti-microbials when treating toxoplasmic encephalitis in mice. The additive role of gamma interferon in the treatment of toxoplasmic encephalitis in combination with conventional antimicrobial agents (pyrimethamine/sulfadiazine) will be assessed in upcoming clinical trials. Studies have also demonstrated that interbelurin 2 enhanced survival and reduced the number of cysts formed in the brains of infected animals.

Increasing attention has surrounded the use of prophylaxis through the administration of antitoxoplasmic agents in attempts to prevent reactivation of latent disease. The two groups targeted include (1) HIV-infected patients who are seropositive for toxoplasmosis and possess CD4 cell counts lower than 200; and (2) heart-lung transplant recipients seropositive for *T. gondii* and receiving a graft from a seronegative donor (2).

Another promising development that may play a role in the prevention of toxoplasmosis is the recent detection of a 23-kilodalton (p23) major antigen secreted by *T. gondii* that may be essential as a component of a future vaccine (66). p23 Antigen appears to be a marker for chronic toxoplasmosis.

Acknowledgments. The authors are grateful to Aletha Arkie for her expert secretarial assistance; to David O. Davis, M.D. and Thomas Dina, M.D. from the Department of Radiology for providing the CT and MRI scans; and to Frank Jannotta, M.D. and Arnold Schwartz, M.D., from the Department of Pathology for the histopathology slides.

References

1. Luft BJ, Remington JS: Toxoplasmic encephalitis. J Infect Dis 1988; 157:1–6.
2. LePort C, Raguin G: Toxoplasmic encephalitis. Curr Opin Infect Dis 1990;3:614–619.
3. McCabe RE, Remington JS: Toxoplasma gondii. In Mandell GL, Douglas G, Bennett JE (eds): Principles and Practice of Infectious Diseases. New York, Churchill Livingstone, 1990, pp. 2090–2103.
4. Hofflin JM, Remington JS: Tissue culture isolation of Toxoplasma from blood of a patient with AIDS. Arch Intern Med 1985;145:925–926.
5. Derouin F, Mazeron MC, Garin YJF: Comparative study of tissue culture and mouse inoculation methods for demonstration of Toxoplasma gondii. J Clin Microbiol 1987;25:1597–1600.
6. Sacks JJ, Roberto RR, Brooks WF: Toxoplasmosis infection associated with new goat's milk. JAMA 1982;248:1728–1732.
7. Luft BJ, Remington JS: Acute toxoplasmic infection among family members of patients with acute lymphadenopathic toxoplasmosis. Arch Intern Med 1984;144:53–56.
8. Neu HC: Toxoplasmosis transmitted at autopsy. JAMA 1967;202:284.
9. Papoz L, Simondon F, Saurin W, Sarmini H: A simple model relevant to toxoplasmosis applied to epidemiologic results in France. Am J Epidemiol 1986;123:154–161.
10. Siegel SE, Lunde MN, Gelderman AH, et al: Transmission of toxoplasmosis by leukocyte transfusion. Blood 1971;37:388–394.
11. Britt RH, Enzmann DR, Remington JS: Intracranial infection in cardiac transplant recipients. Ann Neurol 1981;9:107–119.
12. Suzuki Y, Orellana MA, Schreiber RD, Remington JS: Interferon-gamma: the major mediator of resistance against Toxoplasma gondii. Science 1988;240:516–518.
13. Suzuki Y, Conley FK, Remington JS: Importance of endogenous IFN-gamma for prevention of toxoplasmic encephalitis in mice. J Immunol 1989;143:2045–2050.
14. McCabe RE, Luft BJ, Remington JS: Effect of murine interferon gamma on murine toxoplasmosis. J Infect Dis 1984;150:961–962.
15. Murray HW, Hillman JK, Rubin BY, et al: Patients at risk for AIDS-related opportunistic infections: clinical manifestations and impaired gamma interferon production. N Engl J Med 1985;313:1504–1510.
16. Canessa A, DelBono V, Miletich F, Pistoia V. Serum cytokines in toxoplasmosis: increased levels of interferon-gamma in immunocompetent patients with lymphadenopathy but not in AIDS patients with encephalitis. J Infect Dis 1992;165:1168–70.
17. Wilson CB, Haas JE: Cellular defenses against Toxoplasma gondii in newborns. J Clin Invest 1984;73:1606–1616.
18. Vollmer TL, Waldor MK, Steinman L, Conley FK: Depletion of T4+ lymphocytes with monoclonal antibody reactivates toxoplasmosis in the central nervous system: a model of superinfection in AIDS. J Immunol 1987;138:3737–3744.

19. Israelski DM, Araujo FG, Conley FK, et al: Treatment with anti-L3T4 (CD4) monoclonal antibody reduces the infammatory response in toxoplasmic encephalitis. J Immunol 1989;142:954–958.
20. Sibley LD, Weidner E, Krahenbul JL: Phagosome acidification blocked by intracellular Toxoplasma gondii. Nature 1985;315:416–419.
21. Murray HW, Rubin BY, Carruro SM, et al: Human mononuclear phagocyte antiprotozoal mechanisms: oxygen-dependent vs. oxygen-independent activity against intracellular Toxoplasma gondii. J Immunol 1985;134:1982–1988.
22. Ferguson DJP, Hutchison WM, Petersen E: Tissue cyst rupture in mice chronically infected with Toxoplasma gondii an immunocytochemical and ultrastructural study. Parasitol Res 1989;75:599–603.
23. Suzuki Y, Conley FK, Remington JS: Differences in virulence and development of encephalitis during chronic infection vary with the strain of Toxoplasma gondii. J Infect Dis 1989;159:790–794.
24. Luft BJ, Remington JS: Effect of pregnancy on resistance to Listeria monocytogenes and Toxoplasma gondii: infection in mice. Infect Immun 1982;38:1164–1171.
25. Wreghitt TG, Hakim M, Gray JJ, et al: Toxoplasmosis in heart and lung transplant recipients. J Clin Pathol 1989;42:194–199.
26. Beach PG: Prevalence of antibodies to Toxoplasma gondii in pregnant women in Oregon. J Infect Dis 1979;140:780–783.
27. Desmonts G, Couvreur J: Congenital toxoplasmosis: a prospective study of 378 pregnancies. N Engl J Med 1974;290:1110–1116.
28. Wilson CB, Remington JS, Stagno S, Reynolds DW: Development of adverse sequelae in children born with subclinical congenital Toxoplasma infection. Pediatrics 1980;66:767–774.
29. O'Connor GR. Manifestations and management of ocular toxoplasmosis. Bull NY Acad Med 1974;50:192.
30. Dorfman RF, Remington JS: Value of lymph node biopsy in the diagnosis of acute acquired toxoplasmosis. N Engl J Med 1973;289:878–881.
31. McCabe RE, Brooks RG, Dorfman RF, Remington JS: Clinical spectrum of 107 cases of toxoplasmic lymphadenopathy. Rev Infect Dis 1987; 9:754–774.
32. Wanke C, Tuazon CU, Kovacs A, et al: Toxoplasma encephalitis in patients with acquired immunodeficiency syndrome: Diagnosis and response to therapy. Am J Trop Med Hyg 1987;36:509–516.
33. Levy RM, Bredesen DE: Central nervous system dysfunction in acquired immunodeficiency syndrome. J Acquir Immune Defic Synd 1988;1:41–46.
34. Carrazana EJ, Rossitch E, Samuels MA: Cerebral toxoplasmosis in the acquired immune deficiency syndrome. Clin Neurol Neurosurg 1989; 91:291–299.
35. Haverkos HW: Assessement of therapy for toxoplasmic encephalitis: the TE study group. Am J Med 1987;82:907–914.
36. Navia BM, Petito CK, Gold JWM, et al: Cerebral toxoplasmosis complicating the acquired immune deficency syndrome: clinical and neuropathological findings in 27 patients. Ann neurol 1986;19:224–238.

37. Parke DW, Font DL: Diffuse toxoplasmic retinochoroiditis in a patient with AIDS. Arch Ophthalmol 1986;104:571–575.
38. Luft BJ, Conley FK, Remington JS: Outbreak of central nervous system toxoplasmosis in Western Europe and North America. Lancet 1983; 1:781–784.
39. Catterall JR, Hofflin JM, Remington JS: Pulmonary toxoplasmosis. Am Rev Respir Dis 1986;133:704–711.
40. Nistal M, Santana A, Paneaquia R, Palacios J: Testicular toxoplasmosis in two men with the acquired immunodeficency syndrome. Arch Pathol Lab Med 1986;110:744–746.
41. Farkash AE, Maccabbee PJ, Sher JH: CNS toxoplasmosis in acquired immune deficiency syndrome: a clinical pathological radiographical review of 12 cases. J Neurol Neurosurg Psychiatry 1986;49:744–748.
42. Kupfer MC, Zee CS, Colletti PM, et al: MRI evaluation of AIDS-related encephalopathy: toxoplasmosis vs. lymphoma. Magn Reson Imag 1990;8(1):51–57.
43. Zee CS, Segall HD, Rogers C, et al: MR imaging of cerebral toxoplasmosis: correlation of computed tomography and pathology. J Comput Assist Tomogr 1985;9:797.
44. Levy RM, Mills CM, Posin JP, et al: The efficacy and clinical impact of brain imaging in neurologically symptomatic AIDS patients: a prospective CT/MRI study. J Acquir Immune Defic Syndr 1990;3:461–471.
45. Levy RM, Rosenbloom S, Perrett L: Neuroradiologic findings in the acquired immunodeficiency syndrome: a report of 200 cases. Am J Nucl Radiol 1986;7:833–839.
46. Luft BJ, Brooks RG, Conley FK, et al: Toxoplasmic encephalitis in patients with acquired immune deficiency syndrome. JAMA 1984;252:913–917.
47. Potasman I, Resnick L, Luft BJ, Remington JS. Intrathecal production of antibodies against T. gondii in patients with toxoplasmic encephalitis and AIDS. Ann Intern Med 1988;108:49–51.
48. Weiss LM, Udem SA, Tanowitz H, Wittner M: Western blot analysis of the antibody response of patients with AIDS and Toxoplasma encephalitis: antigenic diversity among Toxoplasma strains. J Infect Dis 1988; 157:7–13.
49. Burg JL, Grover CM, Pouletty P, Boothroyd JC: Direct and sensitive detection of a pathogenic protozoan, Toxoplasma gondii, by polymerase chain reaction. J Clin Microbiol 1989;27:1787–1792.
50. Lebach M, Lebach AM, Nelsing S, et al: Detection of *Toxoplasma gondii* DNA by polymerase chain reaction in cerebrospinal fluid from AIDS patients with cerebral toxoplasmosis. (letter) J Infect Dis 1992;165:982–3.
51. Brooks RG, McCabe RE, Remington JS: Role of serology in the diagnosis of toxoplasmic lymphadenopathy. Rev Infect Dis 1987;9;775–782.
52. Wong B, Gold JM, Brown AE, et al: Central nervous system toxoplasmosis in homosexual men and parenteral drug abusers. Ann Intern Med 1984;100:36–42.
53. Bishburg E, Eng RHK, Slim J, et al: Brain lesions in patients with ac-

quired immunodeficiency syndrome. Arch Intern Med 1989;149:941–943.
54. Tuazon CU: Toxoplasmosis in AIDS patients. J Antimicrob Chemother 1989;23:77–82.
55. LePort C, Raffi F, Matheron S, et al: Treatment of central nervous system toxoplasmosis with pyrimethamine/sulfadiazine combination in 35 patients with the acquired immunodeficiency syndrome: efficacy of long-term continuous therapy. Am J Med 1988;84:94–100.
56. Rolston KV, Hoy J: Role of clindamycin in the treatment of central nervous system toxoplasmosis. Am J Med 1987;83:551–554.
57. LePort C, Bastuji-Garin S, Perronne C, et al: An open study of the pyrimethamine-clindamycin combination in AIDS patients with brain toxoplasmosis. J Infect Dis 1989;160:557–558.
58. Dannemann B, McCutchan A, Israelski D, et al: Treatment of toxoplasmic encephalitis in patients with AIDS. A randomized trial comparing pyrimethamine plus clindamycin to pyrimethamine plus sulfadiazine. Ann Intern Med 1992;116:33–43.
59. Kovacs JA, Allegra CJ, Beaver J, et al: Characterization of de novo folate synthesis in Pneumocystis carinii and Toxoplasma gondii: potential for screening therapeutic agents. J Infect Dis 1989;160:312–320.
60. Araujo FG, Guptill DR, Remington JS: Azithromycin, a macrolide antibiotic with potent activity against Toxoplasma gondii. Antimicrob Agents Chemother 1988;32:755–757.
61. Chang HR, Pechere JF: In vitro effects of four macrolides (Roxithromycin, Spiramycin, Azithromycin (CP-62, 993), and A-56268) on Toxoplasma gondii. Antimicrob Agents Chemother 1988;32:524–529.
62. Chang HR, Jefford CW, Pechere JC: In vitro effects of three new 1, 2, 4 trioxanes (Pentatroxane, Thiahexatroxane, and Hexatroxanone) on Toxoplasma gondii. Antimicrob Agents Chemother 1989;33:1748–1752.
63. Fernandez-Martin J, Leport C, Morlat P, et al: Pyrimethamine-clarithromycin combination for therapy of acute toxoplasma encephalitis in patients with AIDS. Antimicrob Agents Chemother 1991;35:2049–2052.
64. Farthing C, Rendel M, Currie B, Seidlin M. Azithromycin for cerebral toxoplasmosis (letter) Lancet 1992 Feb 15;339(8790):437–8.
65. Huskinson-Mark J, Araujo FG, Remington JS. Evaluation of the effect of drugs on the cyst form of *Toxoplasma gondii*. J Infect Dis 1991;164:170–7.
66. Cesbron-Delauw MF, Guy B, Torpier G, et al. Molecular characterization of a 23-kilodalton major antigen secreted by Toxoplasma gondii. Proc Natl Acad Sci USA 1989;86:7537–7541.

3
Anisakidae and Anisakidosis

Hajime Ishikura, Kokichi Kikuchi, Kazuya Nagasawa, Toshio Ooiwa, Hiroshi Takamiya, Noriyuki Sato, and Kazuo Sugane

During the 1950s, Ishikura noticed in the fishing town of Iwanai, Hokkaido, an intestinal disease that frequently occurred in winter with clinical and histopathologic characteristics different from those of ordinary ileitis terminalis. He reported eight such cases, which occurred within a 2-month period in 1955, as acute ileitis terminalis showing peculiar pathologic features (1). By 1959, thirty cases of this disease had been reported (2). Ishikura, Kikuchi, and their coworkers considered the lesion peculiar because this regional enteritis (or terminal ileitis) caused severe allergic tissue reaction with extensive eosinophilic infiltration. On histologic examination, cross sections of a nematode-like worm were sometimes found, and the specimens were sent to parasitologists for identification. The parasitologists regarded these structures as roundworm larvae, not realizing that they were *Anisakis* larvae. Meanwhile Otsuru et al. (3), based on a description by Beaver (4), warned that visceral larva migrans might occur in Japan.

In 1960 van Thiel et al. (5) found in the Netherlands the larva of a nematode in the intestinal wall of a patient who, after eating raw herring, suffered an acute abdominal syndrome. The larva was identified as *Eusutoma rotundatum*, and it was the first time that an *Anisakis simplex* larva was shown to be involved in larva migrans. In 1962 the disease was named anisakiasis (6), and it is now called anisakidosis. The nematode-like worms found in the Iwanai cases were later identified as *Anisakis simplex* larvae. Thus intestinal anisakidosis was first observed in both the Netherlands and Japan. Since Namiki et al. (7) removed an *Anisakis* larva from the gastric wall under gastric endoscopy in 1968, gastric anisakidosis has been found with increasing frequency in Japan owing to the improvement of gastric endoscopic techniques.

Anisakid larvae other than *Anisakis simplex* and *Pseudoterranova decipiens*, such as *Contracaecum osculatum*, have been found to cause disease in humans. We therefore include in this review dis-

Table 3.1 Occurrence of anisakidosis in the world.

| | *Anisakis simplex* | | *Phocanema* | | | |
Country	Gastric	Intest.	Gastric	Intest.	Unknown	Total
The Netherlands	3	289	0	0	0	292
West Germany	13	61	0	0	7	81
France	19	19	0	0	26	64
USA	12	24	11	2	2	51
Korea	6	3	0	1	1	11
GDR	0	10	0	0	0	10
Belgium	0	9	0	0	0	9
UK	0	2	1	0	5	8
Norway	0	1	0	0	5	6
Poland	0	0	0	0	5	5
Chile	1	0	1	0	3	5
Canada	1	0	2	0	0	3
Sweden	1	2	0	0	0	3
Brazil	0	0	0	0	2	2
Israel	2	0	0	0	0	2
Denmark	0	0	0	0	1	1
Italy	0	0	0	0	1	1
New Zealand	0	1	0	0	0	1
Greenland	0	0	1	0	0	1
Western Samoa	0	0	0	0	1	1
Taiwan	0	0	0	0	1	1
Tahiti	1	0	0	0	0	1
Total	59	421	16	3	60	559

orders caused by *Contracaecum*, as well as *Anisakis* and *Pseudoterranova* larvae. Thus anisakidosis is composed of anisakidosis, pseudoterranovosis, and disorders caused by *Contracaecum* and other vaguely defined anisakid larva. For the nomenclature of fish, we have used the *Dictionary of Japanese Fish Names and Their Foreign Equivalents* published by the Ichthyological Society of Japan.

Epidemiologic Aspects of Anisakidosis

In Japan there have been 11,629 reported cases of gastric, 567 cases of intestinal, and 45 cases of extragastrointestinal anisakidosis, and 335 cases of gastric pseudoterranovosis. Counting another 10 cases infected by unknown larvae, the total becomes 12,586 cases (8). Outside of Japan, there have been 519 cases of anisakidosis in 19 countries, consisting of 39 cases of gastric anisakiosis, 402 of intestinal anisakiosis, 14 of gastric pseudoterranovosis, 2 of intestinal pseudoterranovosis, and 62 of anisakidosis caused by unknown larvae (8). We have summarized the non-Japanese cases in Table 3.1. When the data were compiled the literature of the USSR was not available, nor was that of many Southeast Asian countries. It should be noted, however, that research on Anisakidae started during the 1800s in the USSR, and people often eat raw fish in Southeast Asia, so there may be additional anisakidosis in these countries.

Incidence of Gastric and Intestinal Anisakidosis

In Japan gastric anisakidosis occurs much more frequently than intestinal anisakidosis: 93.0% and 4.4%, respectively. In other countries intestinal anisakidosis occurs in 75.3% and gastric anisakidosis in 10.6%.

In the sea around Japan, more than 166 species of fish are heavily infected by anisakidae larvae. The Japanese eat much more raw fish than do people in American and European countries, and therefore anisakidosis occurs frequently in Japan. The reason the incidence of gastric anisakidosis is much higher than intestinal anisakidosis in Japan appears to be because gastroscopy has been frequently used to visualize and remove worms. Eighty-one percent of the cases suspected as being anisakidosis have been subjected to gastroscopy, so that larvae were found and removed, with the result that only a few larvae reach the intestine. Gastroscopy is not performed as frequently in other countries as in Japan; therefore the incidence of gastric anisakidosis is relatively low.

Table 3.2 Pseudoterranovosis in Japan.

Locality (district)	No. of cases	No. of anisakidosis caused by *Anisa* type 1 larvae	Reference no.
Hokkaido			
Bihoro	11	25	117
Asahikawa	60	127	118
Hakodate	9	19	119
Rest of Hokkaido	59	290	120
Total	139	461	121
Honshu			
Hirosaki	21	Unknown	122
Goshogawara	31	28	123
Toyama	1	Unknown	124
Ishikawa	18	103	120
Fukui	1	113	120
Tokyo	2	24	120
Total	74	$268 + x$	
Kyushu			
Fukuoka	1	857	125
Ooita	1	533	126
Total	2	1,419	
Total no. in Japan	215	$2,219 + x$	

Pseudoterranovosis

The number of gastric pseudoterranovosis cases reported in non-Japanese countries is 16, whereas it is 335 in Japan where anisakidosis is particularly common. Of the 16 non-Japanese cases, 11 were from the United States, 10 cases of which were "tingling throat anisakidosis" (9). In Japan, most of the 335 cases have been from the northern part of Tohoku and Hokkaido. The occurrence of the disease is locally concentrated in Japan (Table 3.2).

There may be two reasons for the endemic occurrence of the disease in Japan. First, peculiar dietary customs prevail in these endemic areas; for example, in the northeastern part of Tohoku district, sashimi of shark, often infected with *Pseudoterranova* larva, is frequently eaten. Second, doctors in the areas where many patients have been treated for this disease are well aware of the disease and skillful in diagnosis.

Intestinal pseudoterranovosis has been infrequently reported: two cases from the United States and one from Korea, but none from Japan. A case-by-case, precise identification of larvae may eventually reveal the presence of the disease in Japan.

Table 3.3 Survey of simultaneous multiple infections in Kyushu Island (1990).

Location (Prefecture)	Total anisa-kidosis cases	Cases of simultaneous multiple infections									
		No. of worms									
		2	3	4	5	6	7	8	9	10	Total
Fukuoka	1018	93	26	5	2	1	2	0	2	0	131
Saga	284	26	7	4	3	0	0	0	0	1	41
Nagasaki		Ongoing investigation										?
Ooita		Ongoing investigation										?
Kumamoto	255	37	10	3	0	1	0	0	0	0	51
Miyazaki	306	32	6	0	0	0	1	0	0	0	39
Kagoshima		Making close investigation										
Okinawa	49	5	2	1	0	0	0	0	0	1	9
Total	1912	193	51	13	5	2	3	0	2	2	271
	(71.2%)	(18.8%)	(4.8%)	(1.9%)	(0.7%)	(1.1%)	(0%)	(0.7%)	(0.7%)	(14.2%)	

Simultaneous Multiple Infections

In the available literature from outside Japan, we found no simultaneous human infection by multiple Anisakidae larvae, whereas in Japan there are many reports of multiple infections (Table 3.3). According to Iino (unpublished data. 1990), of 1,912 acute symptomatic gastric anisakidosis cases that occurred in Kyushu over a 3-year period, 271 cases (14.2%) involved multiple larvae. Iino also indicated that a large number of multiple infection cases have occurred more recently, which may be related with the increase in infection density of fish around Kyushu Province.

Paratenic Hosts as Sources of Human Anisakidosis

The fish species that are the source of infection for human anisakidosis vary somewhat between the northern and the southern Japanese islands. Furthermore, changes in fish species distribution have been occurring, affected by even such political measures as international fishing restrictions. Other factors include changes in the Japan Sea current; changes in the growth of plankton; move/interchange of cold-current and warm-current dolphins in the Japan Sea current; proliferation of dolphins, seals, fur seals, and Steller's sea lions; change in catch; and change in the diet of the Japanese (popularity of eating live fish). These changes have resulted in more heavily paratenic intermediate hosts in the seas around Japan. In addition, shellfish, including oyster and sea squirt, are now suspected as a source of infection.

According to a 1968 survey of 278 cases [10], the sources of infection were chub mackerel (25%), Alaska pollock (19%), arabesque greenling (12.5%), sandfish (12.5%), and squids (62.5%). According to Iwano et al. [11], in their study of 218 cases, the sources of infection were chicken halibut (21.5%), squids (19.1%), tunas (13.8%), arabesque greenling (11.8%), and flat fish (6.5%). Nationwide surveys have revealed that 448 of 1823 patients with anisakidosis clearly stated they ate raw fish before the development of symptoms. In these cases, the sources of infection and the frequencies for the top five species were chicken halibut (21.9%), squids (17.2%), flat fish (12.7%), tunas (6.7%), and codfish (6.5%).

In the north of Japan, many pseudoterranovosis cases break out owing to the habit of eating raw codfish, whereas the outbreak of anisakidosis has been on the decrease. The decrease in anisakidosis is caused by the decreased infection density of *Anisakis simplex* larvae in the paratenic host fish in this area.

In the south of Japan, pseudoterranovosis cases are only rarely found, but many anisakidosis cases (average 1,500 per year) are being currently detected. Infections by *Anisakis* larva are caused by chub mackerel in the Pacific, horse mackerel in the East China Sea, and sardine, with the infection rate being explained by the high larval infection density per individual fish.

An increasing number of patients have insisted that the infection source was shellfish. Hirata et al. (12) reported five patients in Tokyo who were infected by eating mussel (Mytilidae). Omachi et al. (13) reported a case of mussel and another case of ark shell infection. In Hokkaido, larva migrans caused by octopus, oyster, crawfish, lobster, and sea squirt have been reported. We examined 660 shellfish for the presence of anisakidae larvae. Although we could find neither *Anisakis simplex* nor *Pseudoterranova* larva, we did find four *Malacobdella japonica* parasitized in 1 of every 30 *Spisula sachalinensis*. In addition, one *Hysterothylacium* larva was found in the head of a *Paroctopus dofleini* and one *Thysanoessa* (Euphausiidea) from one *Crassotrea gigas*.

Taxonomy of Parasites

The nematodes that cause human anisakidosis are third-stage larvae of *Anisakis simplex* (Rudolphi, 1809) and *A. physeteris* Baylis, 1923. Likewise third-stage larvae of *Pseudoterranova decipiens* (Krabbe, 1878) are responsible for pseudoterranovosis in man. These nematodes all belong to the family Anisakidae of the superfamily Ascaridoidea (14). Based on their final hosts, *A. simplex* and *P. decipiens* are presently called "whaleworm" and "sealworm," respectively, although "herringworm" and "codworm" were previously used for the individual species.

There is a long history of confusion in the taxonomy of species of the genera *Anisakis* Dujardin, 1845 and *Pseudoterranova* Mozgovoy, 1950. Concerning *Anisakis*, Davey (15) reviewed 21 previously described species of this genus and recognized only three species as valid, two of which are *A. simplex* and *A. physeteris*. These two species have ten and three different synonyms, respectively. However, the recovery of two sibling species (*A. simplex* A and *A. simplex* B) within this species in European waters has again complicated the identification of the species (16–19) and further studies are needed to clarify the identification and taxonomy of the *A. simplex* complex based on larval and adult specimens from other areas including the North and South Pacific and the South Atlantic. In addition, electrophoresis studies (20) have demonstrated that *A. physeteris* and the *A. simplex* complex should be assigned to two distinct subgenera, *Skrjabinisakis* and *Anisakis*. Although this assignation was proposed as early as 1951 (21), the subgenus *Skrjabinisakis* had been regarded as unvalid (14). On the other hand, the generic name of *P. decipiens* has constantly been fluctuating so far, especially over the past 50 years. It was originally described as *Ascaris decipiens*, but *Porrocaecum*, *Terranova*, and *Phocanema* have been used as its generic name. Most recently, it has been placed in the *Pseudoterranova* genus (22,23).

Morphology of Parasites

Third-stage larvae of *A. simplex*, *A. physeteris*, and *P. decipiens* have the following morphologic characteristics: three bilobed lips (one dorsal and two subventral), a boring tooth near the ventral margin of the dorsal lip, and an excretory pore opening at the base of the subventral lips. The esophagus consists of two portions: the preventriculus (= muscular espohagus) and the ventriculus (= glandular esophagus), the latter joining with the intestine. The cuticle surface covers the entire body and has transverse shallow grooves with fine longitudinal striations between them.

Third-stage larvae of *A. simplex*, which have been usually descirbed as *Anisakis* type I larvae, are morphologically characterized by a relatively long ventriculus with an oblique ventriculus–intestinal junction and a short, rounded tail with a mucron (24–27). In live worms, the ventriculus is typically seen as a white opaque spot through the transparent body wall (also the case in *P. decipiens*). Morphologic (28) and physicochemical (29) studies have revealed that type I larvae are identical to *A. simplex* larvae, and it has been verified by *in vitro* cultivation of the larvae (30–33). It is interesting to note that there are some differences in average body size of *A. simplex* larvae between collection localities: 28.4×0.49 mm from northern Japanese waters (24) and 17.0×0.41 mm from the East and South China Seas (34). This difference may be caused by sibling species in these waters, the existence of which has been already evident in European waters.

Formerly described *Anisakis* type II larvae have been recognized to be larval *A. physeteris* by hemoglobin (29) and electrophoresis (19,35) analyses. The larvae have a short ventriculus with a horizontal ventriculus–intestinal junction and a long, conical, tapering tail without a mucron. According to Shiraki (24), the body length and width of the larvae from Japanese waters are 25.7×0.61 mm on average.

Third-stage larvae of *P. decipiens* possess a ventriculus partly covered with an intestinal cecum at its posterior edge and a mucron on the posterior extremity (26). They are yellowish-brown, measuring on average 32.6×0.80 mm (24). In Japan, *P. decipiens* larvae have been described as *Terranova* type A larvae (24,26).

Life Cycle of Parasites

The life cycles of *A. simplex*, *A. physeteris*, and *P. decipiens* are indirect and involve free-living larval stages, parasitic larval stages in intermediate and paratenic (or transport) hosts, and parasitic larval and adult stages in marine mammals as final hosts. Although our knowledge on the life cycles of these nematodes has increased considerably in recent years, much information is needed to fully understand them. The life cycle of *A. simplex* has been reviewed by Oshima (36) Shimazu (37), Smith (38), and Nagasawa (39) and that of *P. decipiens* by McClelland et al. (40) and Hafsteinsson and Rizvi (41).

Anisakis simplex

The life cycle of *A. simplex* is as follows (Fig. 3.1). Eggs are passed into seawater with the feces of marine mammals. They have trans-

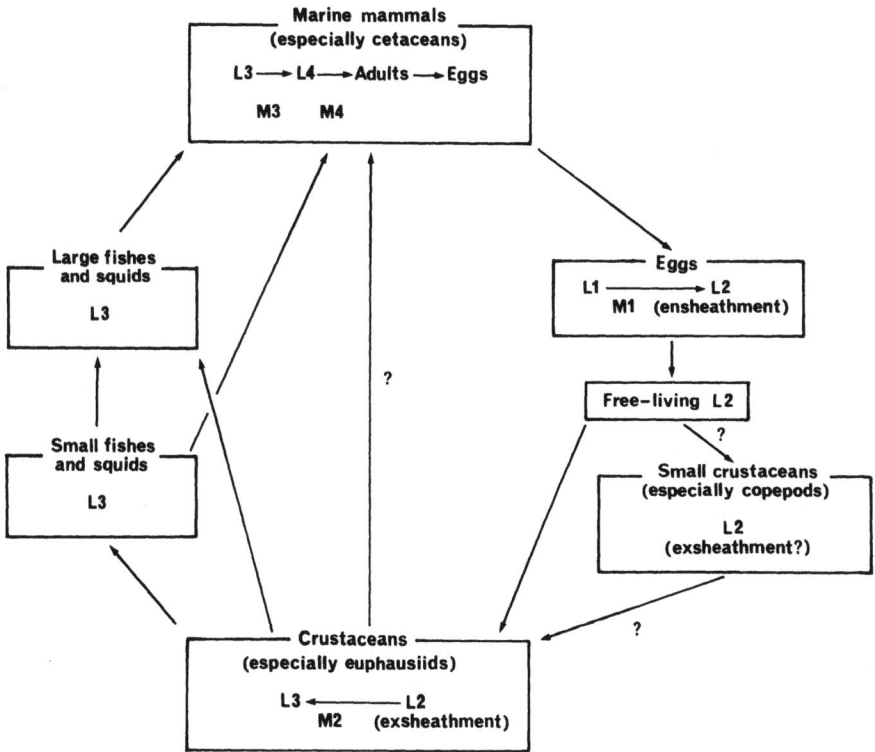

Figure 3.1. Life cycle of *Anisakis simplex* L1, L2, L3, L4 = first-stage, second-stage, third-stage, and fourth-stage larvae: M1, M2, M3, M4 = first, second, third, and fourth moults.

parent and smooth shells and round to oval (27,29), measuring 40 × 50 μm on average (29). The embryonic development proceeds in the sea, and the first moult takes places within the eggs. Hatched free-living larvae, 355 μm average length (29), are ensheathed in the cast cuticle of the moult, being the second stage (27,37,39). They are active in seawater and can survive for 3–4 weeks at 13–18°C and 6–7 weeks at 5–7°C (29).

Although *A. simplex* larvae have been obtained from amphipods (*Caprella septentrionalis*) and decapods (*Hyas araneus*) in the Barents Sea (42) and from caridean prawns (*Pandalus borealis* and *P. kessleri*) in Japanese waters (43), they have also been found in various species of euphausiids in the northern and southern hemispheres: *Thysanoessa raschii* in the Barents Sea (42); *T. inermis, T. longicaudata, T. raschii, Meganyctiphanes norvegica,* and *Nyctiphanes couchi* in the northern North Sea and northeast Altantic (38,44–46); *T. longiceps, Euphausia pacifica,* and *E. nana* in the northern and western North Pacific (47–49); *T. raschii* in the Bering Sea (47); *E. pacifica* in the East China Sea (50); *E. vallentini* in the western Indian Ocean (51); and *N. australis* in New Zealand waters (52). Moreover, experimental exposure of euphausiids (*E. similis* and *E. pacifica*) to hatched larvae of *A. simplex* resulted in successful infections with the larvae (36,53) Thus euphausiids are currently recognized as major intermediate hosts for A. simplex (36–39). Hatched free-living ensheathed second-stage larvae are fed on by euphausiids, in which the larvae exsheath and migrate to the hemocele. The second moult occurs, and the larvae develop to the third stage. They lie free in the hemocele. In addition, Smith (38) has suggested that small crustaceans such as copepods may transfer the second-stage larvae to euphausiids or other crustaceans, although this suggestion has not been verified.

Fishes and squids serve as transport hosts for *A. simplex* (37,39). They acquire infections by feeding on intermediate hosts (mainly euphausiids) that are harboring third-stage larvae. No moult takes place in these hosts. The larvae are usually found encapsulated in their viscera and flesh (Fig. 3.2). If small fish or squid are preyed on by large fish or squid, the larvae are capable of reestablishing in the latter without a moult. In Japanese waters, 122–164 species of fish and one squid species have been reported to be infected with the larvae (36,54).

Marine mammalian final hosts become infected by preying on fish and squid parasitized by third-stage larvae. Although 23–26 cetacean and 11–12 pinniped species have been recroded as hosts of *A. simplex* (15,55), cetaceans appear to be more suitable hosts than pinnipeds because the parasite matures predominantly in the former. *A. simplex* inhabits mainly the stomach, where the third and fourth moults take

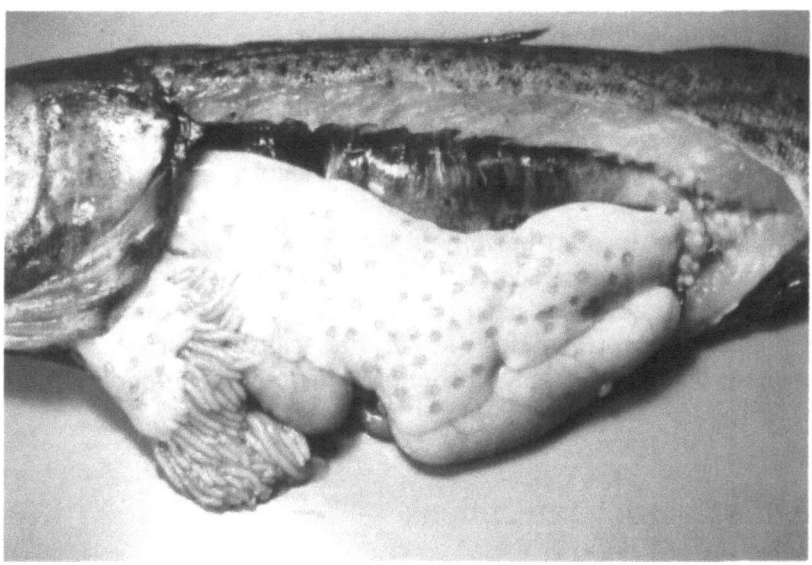

Figure 3.2. *Anisakis simplex* third-stage larvae encapsulated on the liver of the walleye pollock *Theragra chalcogramma* from the western North Pacific off southeast Hokkaido, Japan.

place and it develops to a sexually mature adult. Because third-stage larvae are infective to final hosts, there is a possibility that marine mammals could acquire direct infections by eating crustacean hosts harboring such larvae (38).

Anisakis simplex has been reported principally from colder temperate and polar waters, although it is widely distributed in the oceans and seas of the world (15). Interestingly, however, three species of whales—sei whales (*Balaenoptera borealis*), fin whales (*B. physalus*), and pigmy blue whales (*B. musculus brevicauda*)—from the Antarctic Ocean are free from *A. simplex* (56). This absence of infection is due to the fact that euphausiids in that region (*Euphausia superba*), a dominant prey of these whales, are not infected with the parasite (57).

Anisakis physeteris

Little information is available on the life cycle of *A. physeteris*. According to Hatsushika (58), eggs dissected from the uteri of adult females are mostly spherical or ellipsoidal in shape and measure about 50 μm in diameter. In artificial seawater they hatch within 6 days at 27°C. The larvae are cylindrical and ensheathed, being

0.18–0.24 mm long and 0.02 mm wide including the sheath. They swim actively and survive for 3 weeks in artificial seawater.

No larvae of *A. physeteris* have yet been detected in any marine invertebrates, and thus nothing is known of larval morphology in such hosts. In Japanese waters, third-stage larvae have been recorded from 25–28 species of teleosts (excluding coelacanth) and two species of squid (36,54).

The geographic distribution of *A. physeteris* is worldwide (15). Sperm whales (*Physeter catodon*) are the principal final host of this parasite, although it has been occasionally found in three other species: pygmy sperm whales (*Kogia breviceps*), pilot whales (*Globicephala ventricosus*), and bottlenose whales (*Hyperoodon ampullatus*) (15).

Pseudoterranova decipiens

Eggs are released by gravid females in the stomach of pinnipeds. The egg covering, which consists of the shell and a vitelline membrane, is smooth and bears adhesive properties (59). The eggs are oval and range from 48 to 54 μm in diameter. After being released into seawater, they sink and adhere to the substrate. Embryonic development to ensheathed larva takes place in the eggs. The development and hatching of the eggs are highly dependent on environmental temperatures. They hatch in 8 days at 20°C and 52 days at 5°C but cannot develop at 25°C (60).

Freshly hatched larvae are ensheathed and measure 200–215 × 13–15 μm (60). They are attached to the substrate by their caudal extremities and are active. The survival period of the larvae in seawater varies from 6 days at 17°C to 140 days at 5°C.

Experimental transmission of ensheathed larvae to various copepods has been successful (60), although natural infections in copepods have not yet been detected. The experimentally infected copepods are benthic and epibenthic species belonging to the Harpacticoida and Cyclopoida. When ingested, the larvae exsheath in the gut and penetrate to the hemocoel, where they grow an average of 60% and maximum of 130% in length. However, they do not moult or undergo significant morphologic changes in copepods.

Natural infections of marine invertebrates with *P. decipiens* larvae have been found in polychaets (*Lepidonotus squamatus*) (61), amphipods (*Caprella septentrionalis*) (62), and isopods (*Idothea naglecta*) (63). Moreover, experimental studies (40) have shown that crustanceans (mysids, isopods, amphipods, cumaceans, and decapods), errant polychaets, and molluscs become infected by feeding on copepods haboring the larvae. Thus it is evident that a wide variety of macroinvetebrates can acquire *P. decipiens* larvae. According to McClelland et al. (40), the larvae undergo significant growth and morphologic development in the hemocoele of amphipods. They can reach 2-3 mm in length in 30 days and 7-10 mm in 90 days at 15°C; the large larvae are infective to final hosts, indicating that a fish host is not always necessary in the life cycle of the parasite. On the other hand, the roles of copepods and macroinvertebrates as hosts remain, but there is a suggestion (40) that copepods are transfer or paratenic hosts, not true intermediate hosts. Additionally, McClelland (60) stated that *P. decipiens* larvae, which hatch and invade copepods, may be in the third stage.

Much information is available on the infections of *P. decipiens* larvae in fish, especially in commercially important fish. Hafsteinsson and Rizvi (41) summarized the occurrence of larval *P. decipiens* in cod from the North Atlantic including Canadian and European waters. The larvae are usually found in the flesh (Fig. 3.3). They are transmitted from one species of fish to another through predation (64,65) Small worms may grow to 60 mm in length in fish hosts, although larvae more than 5 mm long are infective to seals (64). In Japanese and Californian waters, the larvae have been recorded from 8-9 and 11 fish species, respectively (36,54,66).

Pinnipeds are final hosts for *P. decipiens*. When infected fish or crustaceans are preyed on by pinnipeds, the larvae escape from the tissues of those hosts and partially embed themselves in the stomach wall (67), The third moult takes places 2-5 days after infection and the fourth moult at 5-15 days (68). The parasites then reach maturity

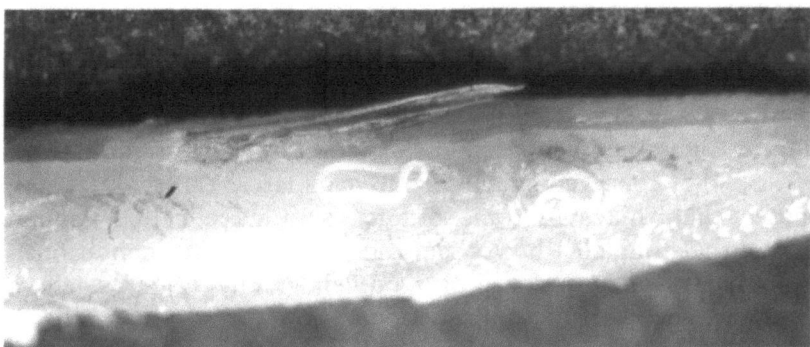

Figure 3.3. *Pseudoterranova decipiens* third-stage larvae found in the muscle of the smelt *Osmerus eperlanus* from the North Sea off the northwest coast of West Germany.

and lay eggs, which are passed into seawater with the host's feces (69).

Clinical Aspects of Human Anisakidosis

Based on the location of the lesion, anisakidosis is divided into gastric, intestinal, and heterologous anisakidosis (ectopic anisakidosis); the latter can be found in various locations of the human body, although it is rare. All the above anisakioses have a fulminant and a mild form, clinically and histopathologically.

Generally, the clinical symptoms of anisakidosis include nausea, vomiting, abdominal and epigastric pain, abdominal dilatation by ascites or intestinal gas filling, movable soft induration, diarrhea followed by normal stool or constipation, and mucous and bloody stool due to intestinal obstruction. Mild pyrexia, leukocytosis, and eosinophilia are seen during the first 3 weeks. An increase in eosinophils is also seen with ascites. Edema, congestion, bleeding, and erosion of the gastric mucosa are observed by gastroscopy. Acute ulcerative changes, edema, and stenosis on ultrasonography and radiography; nematode shadows in the gastrointestinal cavity on roentgenography; and redness, bleeding, white coat adhesion, swelling, thickening of the mesentery, and swelling of lymph nodes as seen by laparoscopy may also be seen. These findings suggest the allergic nature of the disease (70,71). The *Pseudoterranova* larva appears to be less invasive to the digestive tract wall than the *Anisakis simplex* larva. *Pseudoterranova* is more frequently expelled by vomiting than is the *Anisakis simplex* larva.

Vanishing Tumor of the Stomach

In 1950 Gefter et al. (72) presented a paper on vanishing tumor of the lung. Kaye and Stassa (73). found a similar phenomenon in the stomach. The same phenomenon was found by Wright and Matthews in 1971 (74) by roentgenography. Four years later, Yamazaki et al. (75) proposed that this symptom complex should be called vanishing tumor of the stomach when it met the following three conditions: (1) a clear shadow of tumor seen on roentgenography; (2) large size; and (3) disappearance within a short period of time.

In Japan 71 cases of this symptom complex have been reported (Figs. 3.4 through 3.7), with the number of cases still increasing. The original diseases are shown in Table 3.4, as are the locations of the lesions. The largest vanishing tumor was 10.0×7.0 cm and the

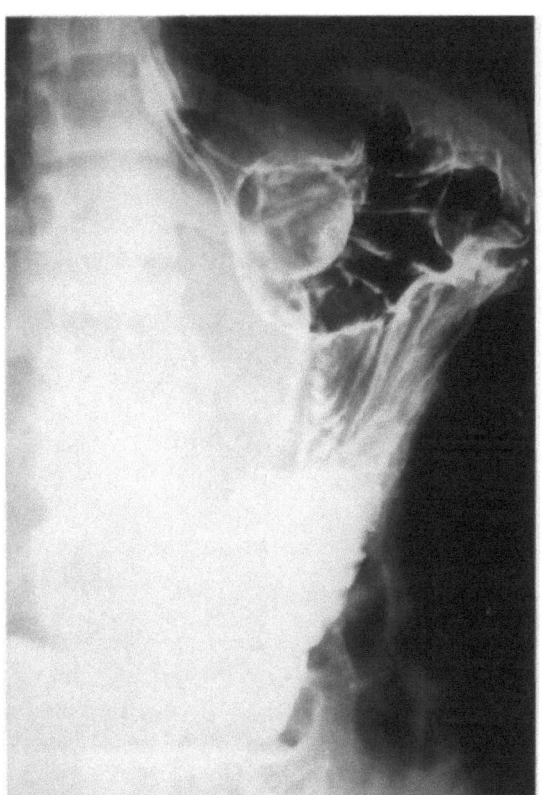

Figure 3.4. Double-contrast barium-filled roentgenographic image (after 22 hours from the onset) showing a large spherical tumor-like shadow in the cardiac-fornix region.

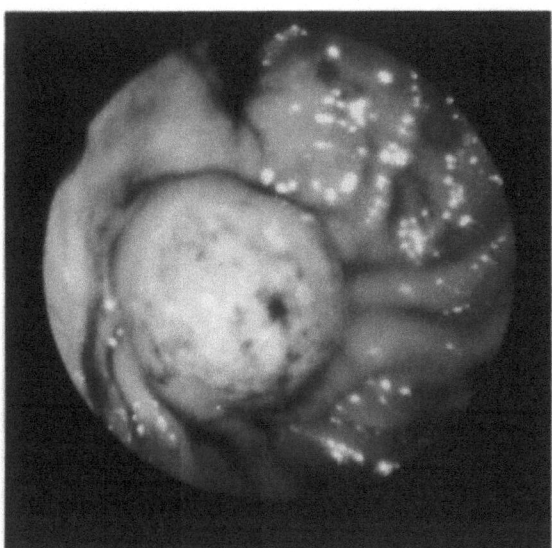

Figure 3.5. Gastroendoscopic findings (after 1 day from the onset) showing a spherical (ca. 4.0 × 4.0 cm) tumor with edematous and hemorrhagic lesions.

Figure 3.6. Roentgenographic image (after 8 days from the onset) of the tumor of Figure 3.4. It has already vanished, and the tissue shows almost normal signs.

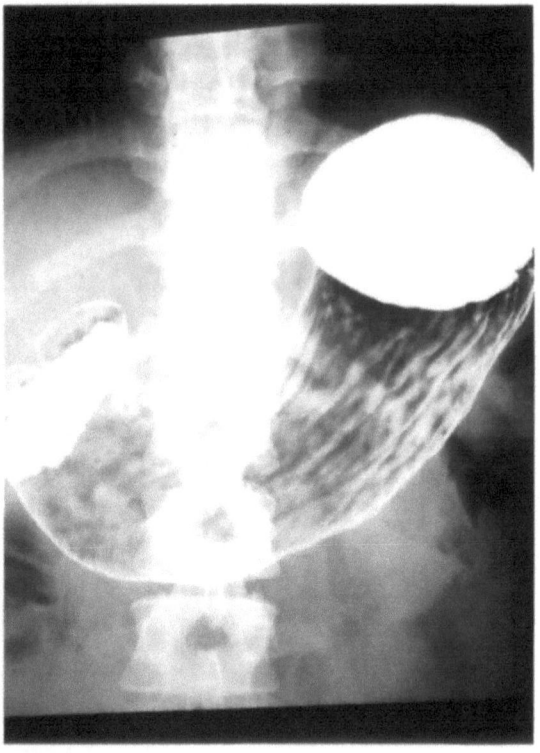

Figure 3.7. Gastroendoscopic findings (after 8 days from the onset) showing that the tumor of Figure 3.4 has vanished, but we see a small area of redness and erosion. These lesions are called AGML phenomena.

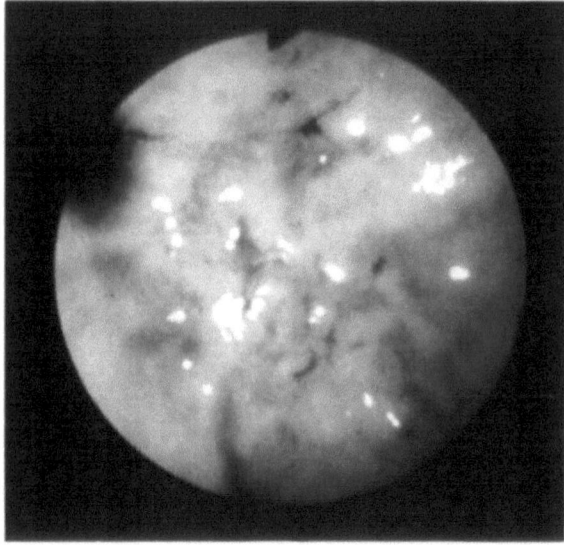

Table 3.4 Analysis of the some information concerning the vanishing tumor.

Parameter	No. of cases
Annual occurrence	
1976	2
1977	4
1978	4
1979	5
1980	8
1981	—
1982	—
1983	9
1984	4
1985	—
1986	4
1987	8
1988	11
1989	8
1990 (September)	4
Total	71
Etiologic investigation: causative diseases	
Anisakiosis	29
Probable anisakiosis	15
Granuloma	1
Lymphadenoma	1
Acute gastritis	1
Hepatoma	1
Gastric phlegmona	1
Edematous duodenal tumor	1
Edematous colonic tumor	1
Edematous pelvic cavity tumor	1
Unknown	19
Total	71
Location of the lesion	
Cardia	9
Fornix	26
Body	8
Angulus	6
Antrum	4
Pylorus	3
Duodenum	1
Colon	1
Pelvic cavity	1
Unknown	12
Total	71

Table 3.5 Size of the vanishing tumor.

Size of tumor	No. of cases
Giant size	7
Fist size	3
Henegg size	3
Giant rugae	1
10.0 × 7.0	4
7.0 × 4.0	2
6.0 × 4.0	2
6.0 × 5.0	1
6.0 × 4.5	1
5.0 × 5.0	1
5.0 × 3.0	1
4.5 × 4.5	2
4.0 × 4.0	1
3.5 × 3.5	2
3.0 × 3.0	2
3.0 × 2.8	1
Thumb size	1
2.5 × 2.1	1
2.0 × 2.0	1
1.5 × 1.5	1
Not described	34
Total	71

Table 3.6 Interval between first examination and disappearance of tumor.

Causative disease	No. of cases	Interval of disappearance (mean)
Anisakiasis	29	
Larva extracted	16	15.7 days
Seroimmuno diagnosis	7*	65.0 days
Not described	6	—
Probable anisakidosis	15	12.6 days
Lymphadenoma	1	7.0 days
Acute gastritis[a]	1	4.0 months
Granuloma	1	4.5 months
Hematoma	1	3.5 years
Gastric phlegmon	1	—
Duodenal vanishing tumor	1	15.4 days
Colonic vanishing tumor	1	1.5 years
Vanishing tumor of pelvic cavity	1	17.0 days
Unknown	19	—
Total	71	

[a] This report had been entitled "Vanishing tumor" by the author, but we think it was the "submucosal tumor" caused by chronic anisakiosis.

Table 3.7 Species of raw fishes and squids eaten by patients before vanishing tumor onset.

Species of paratenic host	No. of cases
Scomber japonicus (including *Shimesaba*, n = 6)	17
Trachurus japonicus	3
Todarodes pacificus	2
Engraulis japonicus	1
Takifugu porphyreus	1
Coryphaena hippurus	1
Pheuronectide	1
Culpea pallasi	1
Hippoglossus stenolepis	1
Thunnus thynnus	1
Paralichthys olivaceus	1
Sashimi	1
Not described	40
Total	71

smallest one was 1.5 × 1.5 cm (Table 3.5). Lesions larger than 4.0 × 4.0 cm mostly originated from anisakidosis. Table 3.6 shows the interval from first appearance to the disappearance of the vanishing tumor on roentgenography in 71 cases, 29 of which were anisakidosis. Table 3.7 shows the kinds of fish involved.

Acute Gastric Mucosal Lesion

The acute gastric mucosal lesion (AGML) was proposed by Katz and Siegel in 1968 (76). Based on endoscopic observation, it is a symptom complex that includes erosion, ulceration, and a hemorrhagic lesion on the gastric mucosa. In Japan, Kawai gave the name (acute gastric lesion) (AGL) to a similar mucosal change in 1973 (77). The etiology of AGML is not completely understood, but anisakidosis is obviously one of the causative diseases (Table 3.8; Fig. 3.8).

Intraperitoneal Anisakidosis

Anisakidae larvae in intraperitoneal anisakidosis can be found in the peritoneal cavity or subserosal layer of the digestive tract (78,79). Ishikura et al. (80) first reported a case where an *Anisakis simplex* larva was alive in a small localized cyst on the serosal surface of the ileum, causing intensive edema. As of 1990 there were 45 cases of intraperitoneal anisakidosis reported in Japan. Tiny perforations were demon-

Table 3.8 Etiology of AGML.

	\multicolumn{7}{c}{Frequency of occurrence}	
				Reference no.				
	110 (%)	111 (no.)	112 (%)	113 [a]	114 (no.)	115 (%)	116 (no.)	Remarks
Neurologic stress	24.7	32				32.8		Emotional stress
Post surgery	4.3							
Brain operation					26			
Other severe disease	3.9				19			
Trauma or burns	3.5				14			
Post endoscopic exam.	1.3	240	0.8					
Alcoholic stress	0.9	23	0.4				17	Alcohol consumption
Drug-induced		8			16	25.5		[b]
Food-induced		3				14.6		Garlic, pepper
Anisakidae larvae		7	3				1	
Another cause	1.7				9	7.3		
Not clear	22.1	107			2	19.8		
No. of case	23.1	420	132	17	86	192		

[a] Real number of cases was not described.
[b] Drug-ingestion-associated AGML (Harada K.)
Nonsteroidal antiinflammatory drugs 45.0%
Steroidal hormone 15.2
Antibiotic drugs 14.9
Anticancer drugs 8.4
Oral antidiabetic drugs 7.4
Other drugs 9.1
100% (139 cases)

Figure 3.8. Relation between anisakidosis and vanishing tumor with AGML.

Fundamental lesion or cause of change			Name of disease
Infiltration of lympatic cells			Lymph adenoma
Probable AGL			Acute gastritis
Vascular lesion	Vanishing tumor		Hematoma of the stomach
Chronic granulomatous infiltration			Nonspecific granuloma
Infection of gastric wall			Gastric phlegmon
Edematous change		Anisakiosis	Fulminant form of gastric anisakiosis (second infection)
Hyperemia or bleeding			First infection with larvae (or antibody negative)
Erosion			Acute allergic gastritis
Ulcer			
			Acute gastritis induced by arteriosclerosis
Arteriosclerotic change (Harada)			Acute neurogenic gastritis
Neurologic stress			Acute gastritis induced by trauma or burns
Another stress		AGML	Acute gastritis
1. Trauma & burns			Acute postoperative gastritis
2. Postsurgery			Acute gastritis induced by endoscopic examination
3. Gastroendoscopy			
4. Severe chronic diseases			
Alcohol induced			Acute gastritis after severe disease
Drug induced			Acute alcoholic gastritis
Diet induced			Drug induced acute gastritis
			Diet induced acute gastritis
Unknown			Unknown

Table 3.9 Location of *Anisakis* larvae in cases of extragastrointestinal anisakidosis in Japan.

Location of larvae	No. of cases
Abdominal cavity	22
Abdominal wall	2
Large omentum	7
Mesentery	7
Liver	2
Pancreas	1
Ovary	2
Lymph node	1
Subcutaneous tissue	2
Mucous membrane of oral cavity	3
Mucous membrane of pharynx	1
Mucous membrane of esophagus	24
Pleural cavity	1
Hernial sac	1
Tongue	1
Uvula	1
Tingling throat anisakidosis	10
Duodenum	29
Total	117

strated in some intraperitoneal anisakidosis cases, implying that precautions against peritonitis should be taken.

Extragastrointestinal Anisakidosis

Extragastrointestinal anisakidosis is also called heterologous or ectopic anisakidosis. There have been 117 reported cases of ectopic anisakidosis. As shown in Table 3.9, larvae were found at various unexpected sites. Ectopic anisakidosis is usually associated with mild clinical manifestations, and is sometimes found accidentally during surgical operations for unrelated disorders.

Prevention of Larval Anisakid Infection

Since ancient times, the Japanese have eaten both sashimi (sliced raw fish) with seasonings of wasabi (Japanese horseradish) and ikasashi (sliced raw squid) with shoga (ginger plant) with a little soy sauce. We and others have examined whether these seasonings as well as sake serve to prevent infection (81).

In a mortality test of 30 larvae, half of them died within 20 minutes and most within 50 minutes in sake (16% alcohol). In human gastric fluid, half of them survived for 135 minutes and six for as long as 240 minutes. In human gastric fluid diluted by an equal quantity of sake, half died within 40 minutes and all within 120 minutes. When gastric fluid was diluted threefold with sake, 20 died within 20 minutes, but three survived for 240 minutes. In a mixture of human gastric fluid and an equal quantity of wasabi solution (100 mg/ml), the mortality of larva was reduced. A mixture of equal quantities of human gastric fluid, wasabi, and sake was slightly more effective than a mixture of equal quantities of human gastric fluid and wasabi. When 5 ml of sake, 0.5 g of wasabi, and 30 larvae were simultaneously consumed, the larvae's ability to invade the rabbit stomach wall was more prominent than that of the control. From these results, it seems that when sake and sashimi are simultaneously consumed, in order to avoid larval invasion, the sake/gastric fluid ratio should be 1:1. On the other hand, it is conceivable that wasabi might not have a prominent preventive action against anisakidosis.

Yamada (82) reported that soy sauce plus horseradish powder, soy sauce alone, and soy sauce plus vinegar, in this order, hinder larval activities in vitro, and that the larvae lost their tissue-invading ability in soy sauce plus horseradish powder in 10 minutes, in soy sauce in 2 hours, and in vinegar in 5 hours in vitro. However, in the cyst form in fish, the larvae maintained their tissue-invading power for 2 hours even in soy sauce plus horseradish powder.

Kasuya et al. (83,84) reported that a 0.9% saline extract from Perilla frutescens (gingerol, shougaol) was able to kill Anisakis simplex larvae. In order to find out what substance was effective, they extracted the substances with methanol and broke it down into fractions 1–4. Fraction 1 was found to be the most effective, killing all the larvae within 24 hours.

Murata and Yasuda (85) examined the effect of Chinese medicine commonly sold on the market on the movement of Anisakis simplex larvae. Anchusan, heiisan and two other drugs had prominent suppressive effects, and they found that anethole had the most effective ingredients. In an anethole solution (10.2 mg/dl) the upper walls of the larval esophagus ruptured.

Pathology

The pathology of anisakidosis was first reported in 1965 by Tsuji et al. (86) without knowing its pathogenesis. Ishikura et al. (80) also described clinicopathologic features of intestinal anisakidosis in Japan. Kojima et al. (87) proposed the pathologic classification of gastrointes-

Table 3.10 Pathologic classification of anisakidosis.

Gastric anisakidosis
Intestinal anisakidosis
Extragastrointestinal anisakidosis

Acute type or fulminant form
Chronic type or mild form
   First stage: phlegmon type
   Second stage: abscess type
   Third stage: abscess-granuloma type
   Fourth stage: granuloma type

tinal anisakidosis, and since then several studies of the pathology of anisakidosis have been reported (88–91). Anisakidosis can be classified pathologically, as shown in Table 3.10, according to the location of the disease, stages, pathologic manifestations, and immunologic status.

Location of Disease and Pathology Examination

Anisakis simplex or *Pseudoterranova decipiens* larvae are found in the mouth and the pharyngeal or laryngeal area (Fig. 3.9), sometimes penetrating the mucous membrane and causing the "tingling throat" syndrome. Most of the larvae found there are considered to be from the stomach. Penetration of the worm into the esophageal wall is also reported. Some of the ingested larvae pass through the digestive tract and are evacuated dead, rarely alive, from the anus. Other larvae penetrate the wall of the gastrointestinal tract, inducing symptoms of anisakidosis. These larvae, then, either die, remaining in the wall or passing through the wall and entering the abdominal cavity, producing granulomatous lesion in the omentum, mesentery, or even the abdominal wall. Migrations of the worm into the liver, pancreas, ovary, or pleural cavity are rarely reported.

Of 14,000 reported cases of anisakidosis, gastric anisakidosis was seen in roughly 90% and intestinal anisakidosis in 10%. The high frequency of reported cases of gastric anisakidosis is due to the many successful diagnoses by gastroscopy in Japan.

The location of anisakidosis (i.e., site of larval penetration in the stomach) is most frequently at the fundus(10). If the stomach is divided into four parts—anterior wall, lesser curvature, posterior wall, and greater curvature—the incidence of penetration is highest in the greater curvature (92).

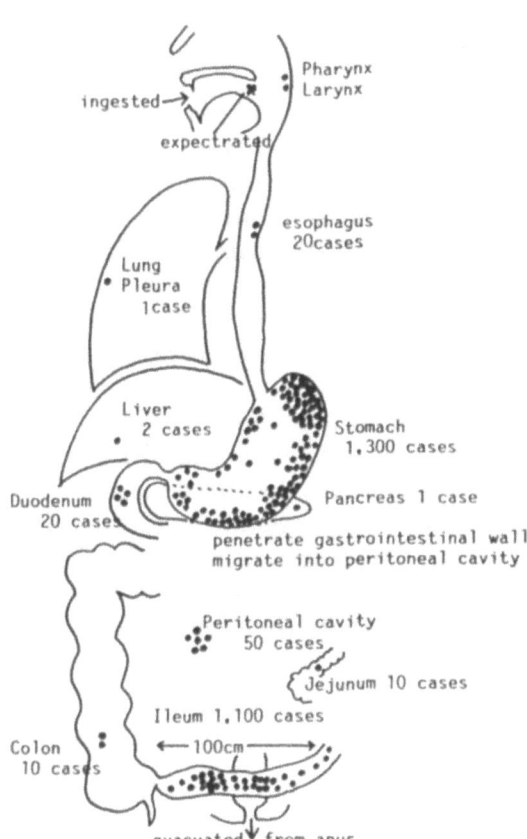

Figure 3.9. Location and frequency of anisakidosis reported up to 1990.

Among intestinal cases, duodenal, jejunal, and colonic anisakidosis have been rare—only about 10, 20, and 10 cases, respectively. Almost all intestinal anisakidosis infestations are concentrated in the terminal ileum between Bauhin's valve and its 100 cm oral from it. The preference of occurrence of anisakidosis in the stomach and terminal ileum might be due to retention of the worm in the areas by the barrier of the pylorus and Bauhin's valve.

Anisakidosis in the abdominal cavity has been reported in only about 50 cases (less frequently than expected) probably bacause in most cases extragastrointestinal anisakidosis is the primary infection, to which the host tends to react only slightly and is not treated surgically.

The clinicopathologic feature of gastric and intestinal anisakidosis is different. Pathologic features are divided into the acute stage (fulminant form) and the chronic stage (mild form). Gastric anisakidosis is mostly diagnosed in the acute stage by gastroscopy. *Anisakis* larvae that are partially penetrating the gastric wall can be detected and easily removed endoscopically during the acute stage. Therefore surgical specimens of the fulminant form of gastric anisakidosis can rarely be obtained, and pathologic examination is done on biopsy specimens in only a few cases (93). Chronic granulomatous lesions in the stomach by *Anisakis* larvae are sometimes resected as tumorous lesions and examined pathologically. Unlike gastric anisakidosis, endoscopic diagnosis of intestinal anisakidosis is difficult because of technical problems. Lesions of intestinal anisakidosis have been often resected, being misdiagnosed as ileus caused by other diseases. Advances in the diagnostic techniques for anisakidosis, such as roentgenography, ultrasonography, abdominography, and immunologic examinations, make it possible to diagnose acute intestinal anisakidosis without surgery and pathologic examination. However, localized chronic granulomatous or abscess-forming lesions are sometimes resected and examined pathologically. In general, pathologic changes by *Pseudoterranova decipiens* larva infection are almost the same as those by *Anisakis simplex* larva infection, except that *Paseudoterranova decipiens* rarely cause any intestinal lesions.

Macroscopic Findings

Acute Gastric Anisakidosis

Namiki et al. (7) first observed *Anisakis* larvae penetrating the gastric wall of the patients with acute gastric symptoms. Using urgent examinations, they removed the worm endoscopically with biopsy forceps and cured the patients.

Endoscopically, penetrating *Anisakis* larvae have a thin, string-like appearance with a milk-white color. The *Pseudoterranova* larva is larger, broader, and yellow or yellowish-brown (93). During the early phase of penetration, the larva moves actively, exhibiting a fast, winding, coiled or sigmoid form. Later, the larva moves slowly and stretches loosely.

The site of penetration is usually edematous accompanied by hyperemia of various degree, depending on the time period of infection and probably the intensity of immunity against the larva. Erosion and bleeding are observed in some cases but to a lesser extent than the fulminant intestinal form. Sometimes the larva is observed in the gastric lumen without penetrating the stomach wall. The gastric mucosa, other than the site of larva penetration, often shows broad edematous changes. Multiple spotted areas of bleeding and

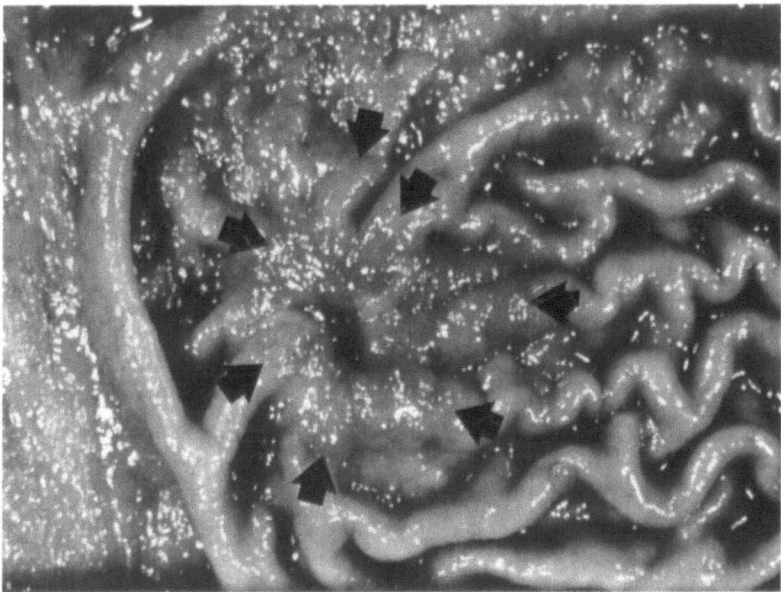

Figure 3.10. Chronic gastric anisakidosis resected. Edema and hyperemia of the mucous membrane (arrows) with central ulceration.

small erosions are also found and are suspected to be induced by the wandering of a larva or by migration of more than one larva at the same time. In some cases, thickening of the longitudinal gastric mucosal folds with edematous changes as caused by penetration of the larva.

Chronic Gastric Anisakidosis

In the affected area, erosion or ulcer with distinct edema on the surface of the localized induration or tumor-like nodule are observed in some cases, which may be misdiagnosed as gastric cancer or gastric ulcer (Fig. 3.10). When the mucous membrane has broad edematous changes, it may be misdiagnosed as Borrman type 4 gastric cancer. When the larva penetrates the gastric wall, the serosa is densely scarred and indurated with adhesion of the major omentum.

Acute Intestinal Anisakidosis

Intestinal anisakidosis, unlike gastric anisakidosis, cannot be diagnosed by endoscopy, and the definitive diagnosis is made only by

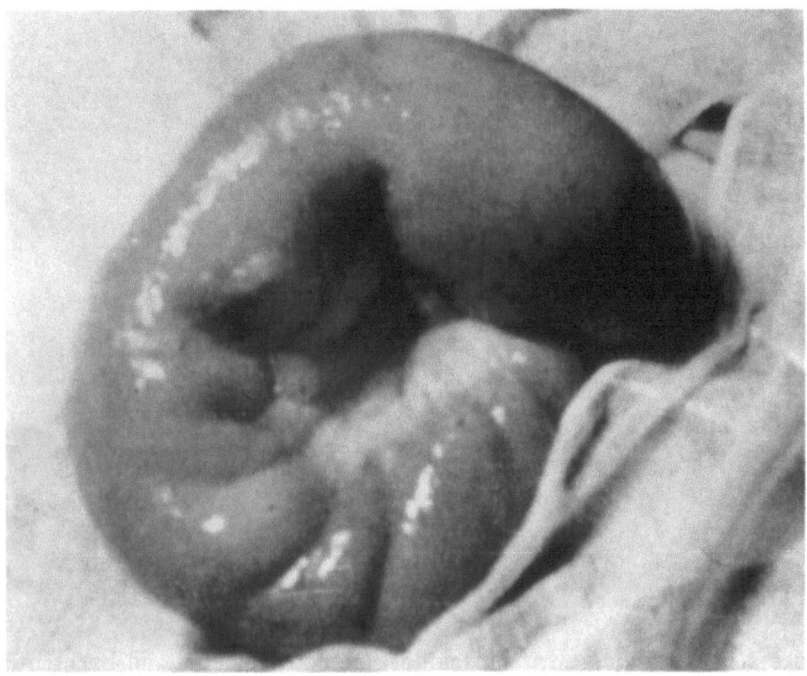

Figure 3.11. Macroscopic appearance of acute intestinal anisakidosis at surgical operation. Note the severe edema of the intestinal wall and marked dilatation of the intestinal lumen.

identifying the larva in the surgically resected specimen or by histologic examination of the specimen. Macroscopic findings of the peritoneal cavity or serosa of the intestine are obtained mainly by careful observation during the operation.

About 300–500 ml of ascites fluid was found in all 19 cases reported by Ishikura (10), although a few cases without ascites has been reported in the early stage of acute intestinal anisakidosis. The appearance of ascites is characteristic: transparent or yellowish translucent serum-like fluid. Ascites may contain free worms migrating through the intestinal wall. Reacted cells are characterized by more than 30% eosinophils. Bacteria are detected in some cases.

The appearance of the peritoneal surface is thin and transparent, which is characteristic compared with other intestinal diseases. The greater omentum shows mild edema, hyperemia, and thickening without necrosis, always accompanied by suppurative inflammation. Fibrous or fibrinous adhesions are present and are usually easily dissected. A soft tumor may be found at the site of adhesion, and the worm may be present in this tumor-like mass. Multiple adhesive

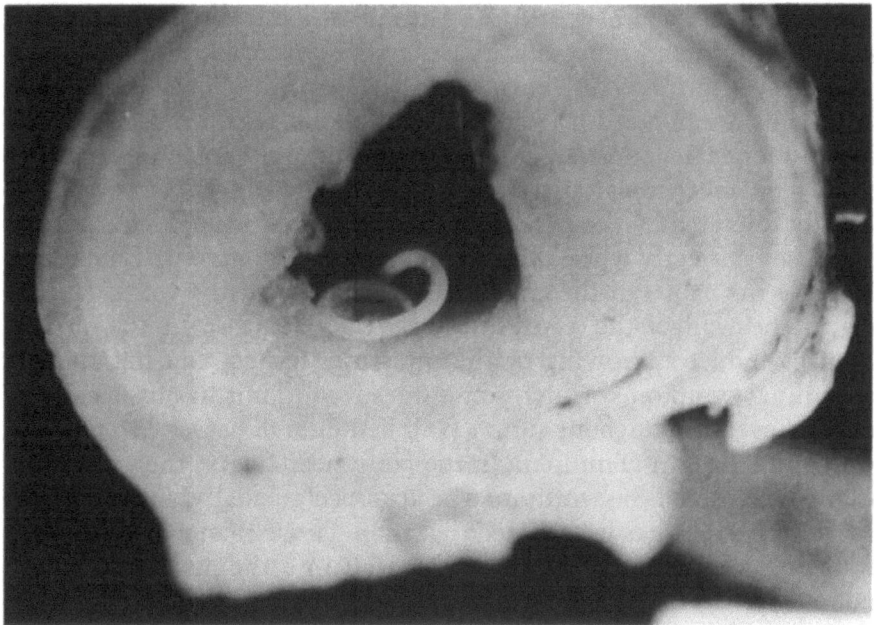

Figure 3.12. Cut surface of a resected ileum with acute intestinal anisakidosis. Note the striking thickening of the intestinal wall, with an *Anisakis* larva penetrating the wall.

lesions, so-called skip lesions, possibly including migrated larva may be found (94). Local mesentery is diffusely edematous in most cases. Local mesenteric lymph nodes are swollen, sometimes larger than the size of thumb tip.

Inflammatory changes in the intestine are concentrated in the terminal ileum and are strictly localized, with severe edema that leads to obstruction of the site of the lesion and proxymal dilatation of the intestinal lumen. The intestinal lesion is palpated as a small lump or cylinder (Figs. 3.11 and 3.12), leading to intestinal obstruction resulting in ileus. The serosa of the affected lesion shows dilatation of the lymphatic duct, hyperemia, petechiae, edema, cloudy swelling, and sometimes necrosis and fibrin clots. Rarely, larvae that have penetrated the wall can be observed moving on the surface of the serosa.

Chronic Intestinal Anisakidosis

In contrast to acute intestinal anisakidosis, which shows fulminant inflammatory symptoms and must be diagnosed before one considers surgery, the chronic form of the disease produces no or mild symptoms. Many cases of this type have been found accidentally on laparotomy for other diseases (10). As a result of clinical observation and animal experiment, most of the cases of this type are considered to be caused by a primary infection with the worm, whereas fulminant symptoms are induced by reinfection by the larva in sensitized hosts as an allergic reaction (10,95). With the primary infection, pain is caused by mechanical irritation by the migrating larvae, which is usually tolerated by the patients, who neither visit the doctor nor are diagnosed correctly when they are examined. Palliative treatment can decrease the pain, but the graulomatous changes by the worm migrating in the wall of the gastrointestinal tract increase in size. These changes induce narrowing of the intestinal lumen, causing chronic abdominal symptoms, which sometimes lead to misdiagnosis as a benign or even a malignant tumor (96). If the larva passes through the wall and creates a granuloma in the peritoneal cavity, the symptoms may be slight and the worm may be absorbed gradually, inducing sensitization (96). It has been noted that more larvae than expected penetrate and pass through the intestinal wall into the peritoneal cavity not only in experimental animals but in humans as well (97). Reinfection occurs when one eats infected raw fish after the primary sensitization. Prompt and severe immunologic reaction at the site of worm penetration causes acute and fulminant symptoms.

Histologic Findings

Acute Gastric Anisakidosis

Almost all the cases of this type have been easily cured by endoscopically removing the worm that has penetrated the gastric wall. Therefore pathologic study of this infestation has become rare. Yazaki and Namiki (93) examined biopsy specimens of these lesions after removing the worm form the stomach. They reported one case of severe eosinophilic infiltration, four cases of submucosal edema, and four cases of inflammatory infiltration among nine patients examined. This variety of histologic appearance might be attributed to the fact that biopsy specimens are usually small samples and that the acute cases of worm penetration include both primary and repeated infections.

Surgical specimens of this type were obtained during the early days when gastroscopy was not available, and they were misdiagnosed as other gastric diseases, or when gastrectomy was carried out for other gastric diseases and anisakidosis coexisted incidentally. In the histologic section, the cut surface of an *Anisakis* larva can be found often

Figure 3.13. Histology of intestinal anisakidosis. Note the cut surface of a larva with marked edema and eosinophil infiltration in the submucosa.

in the submucosal layer (Fig. 3.13). If one can recognize the characteristic Renette cells and the lateral cord in the cut surface of the worm, the diagnosis of anisakidosis is definite. Cross section morphology of the *Pseudoterranova decipiens* larva is characterized by the existence of intestinal cecum on the level of the larval stomach. Even though the *Anisakis* larvae are not found in any one area of the phlegmonous wall, they are occasionally detected by searching an area of several square centimeters. The larva itself is usually intact and clearly visible, preserving its form. It is surrounded by massive neutrophils and eosinophils with some necrosis of the surrounding tissues. In the area of the larva, arteritis and phlebitis with marked inflammatory cell infiltration and fibrinous exudation are found (Fig. 3.14). The gastric wall shows severe thickening with extensive edema and massive eosinophilia mixed with neutrophil, macrophage, and lymphocyte infiltration. Muscle layer and serosa are also involved in the phlegmonous inflammation.

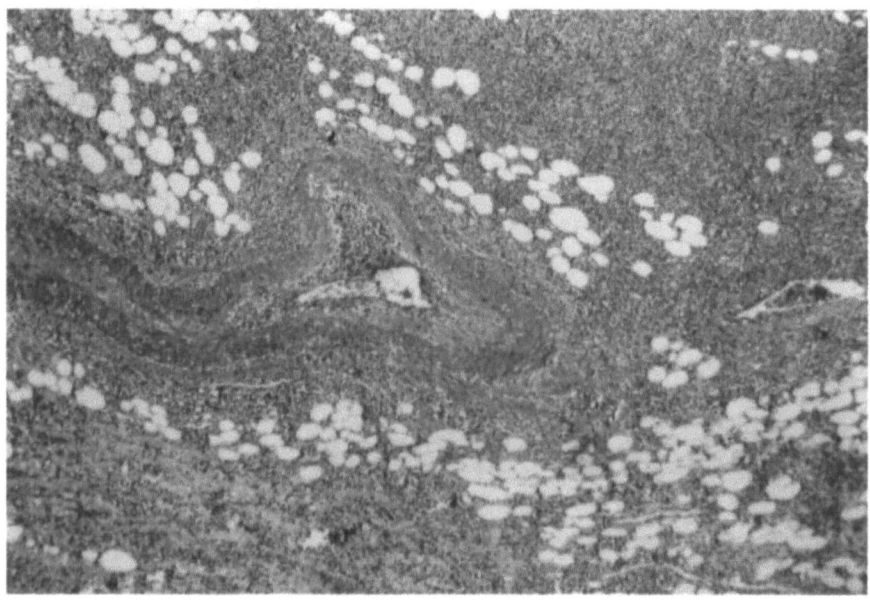

Figure 3.14. Vasculitis of intestinal anisakidosis with marked neutrophil infiltration.

Chronic Gastric Anisakidosis

Kojima et al. (87) classified gastrointestinal anisakidosis histopathologically as four types or stages: first stage, phlegmone type; second stage, abscess type; third stage, abscess-granuloma type; foruth stage, granuloma type.

First Stage—Phlegmonous Type

The phlegmonous stage is induced mainly by an immunologic reaction followed by reinfection with the larva; it is presumably an Arthus-type reaction, as suggested by animal experiments. This type of the disease is encountered less often in the stomach than the intestine. The histologic appearance is almost the same as that of acute gastric anisakidosis described above.

Second Stage—Abscess Type

A marked abscess, with numerous eosinophils, neutrophils, macrophages, and lymphocytes, is found around the slightly degenerating larva or its debris. The degenerating larva is usually surrounded by necrotic tissue at the inner layer of the granuloma, with infiltration of eosinophils into the larva. A small amount of granulomatous tissue surrounding the abscess is infiltrated by eosinophils, macrophages, and lymphocytes with edema and fibrin exudation or fibrinoid degeneration. The muscle layer, as well as the serosa, has the same appearance, though the intensity is varied depending on the cases. The experimental results of Kikuchi et al. (95) and Kojima et al. (87) showed that this phlegmonous reaction was caused by an immunologic reaction to *Anisakis* larva antigen, especially excreted materials, secretions, or visceral antigens, These authors called this type of reaction an "excervation" reaction.

Third Stage—Abscess-Granuloma Type

The abscess is reduced after more than 6 months from the time of laval infection. Degenerated materials or debris of the larva are often found in the center of an abscess surrounded by granulation tissue. The reacting cell infiltration, including eosinophils, is less intensive than that during the first and second stages. The degenerating larva is invaded by the eosinophils and sometimes by the epithelioid cells or foreign body giant cells. Lymphocyte infiltration (instead of eosinophilic infiltration) becomes dominant at this stage.

Fourth Stage—Granuloma Type

Larval debris is surrounded by a small amout of abscess or granulomatous tissue with fibrosis, foreign body giant cells, lymphocytes, and a few eosinophils. This type of lesion is occasionally observed in advanced and long-standing cases of anisakidosis. It may be due to the delay of diagnosis of chronic mild anisakidosis. The degree of destruction of the larva is dependent on the time after infection. Conversely, it is possible that the grade of larval destruction indicates the period of larval infection.

Shiraki (97) compared the pathologic characteristics of anisakidosis and the degree of preservation of the *Anisakis* larva itself. Of the four types of pathologic classification, the best preservation of the larva is seen in the phlegmonous type, where it is occasionally alive. On lesions of the abscess type, the larva is found with the vital cuticle or muscle cells (or both); in the abscess-granulomatous type lesion, these is intense degeneration of all components of the worm. Saeki et al. (98) reported that the grade of destruction of the larva approximately paralleled the degree of immune status of the rabbit that had been subcutaneously and intramuscularly immunized with the larva. From these results, they concluded that immunity is closely related to host reactivity and the degree of larva degeneration, and thus the histologic features. Histopathologic findings and the results of animal experiments suggest that the disease is composed of allergic reac-

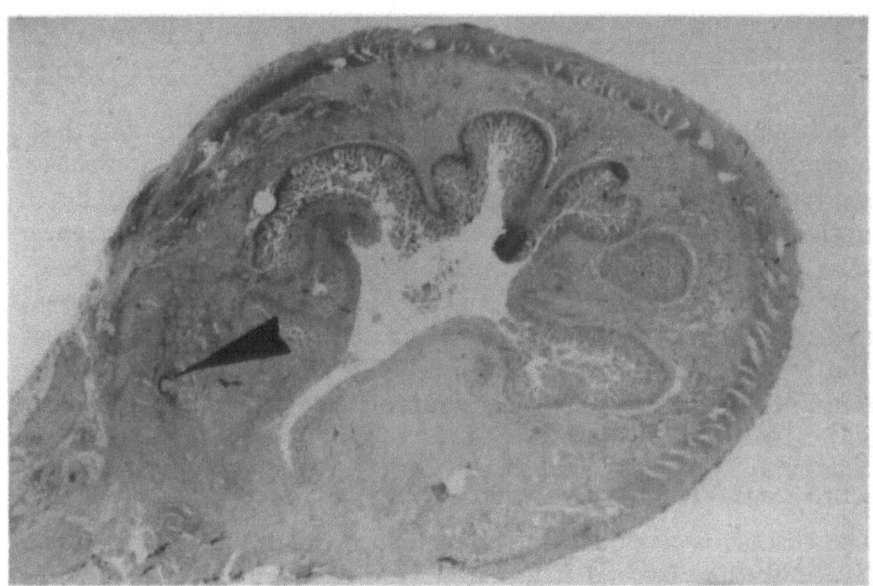

Figure 3.15. Histology of acute intestinal anisakidosis. Note the marked edema of the intestinal wall and narrowing of the intestinal lumen, and the cut surface of an *Anisakis* larva (arrowhead) in the submucosa.

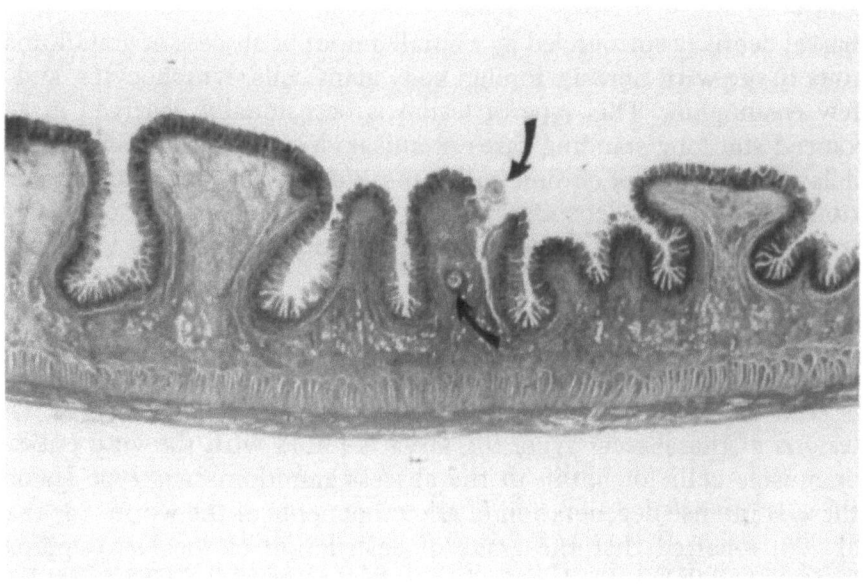

Figure 3.16. *Anisakis* larva is stabbing into the intestinal wall. Two cross sections of one larva can be seen; the upper one is a section at the level of the larval intestine, and the lower one is a section of the larval stomach. Striking edema and marked inflammatory cell infiltration are recognized.

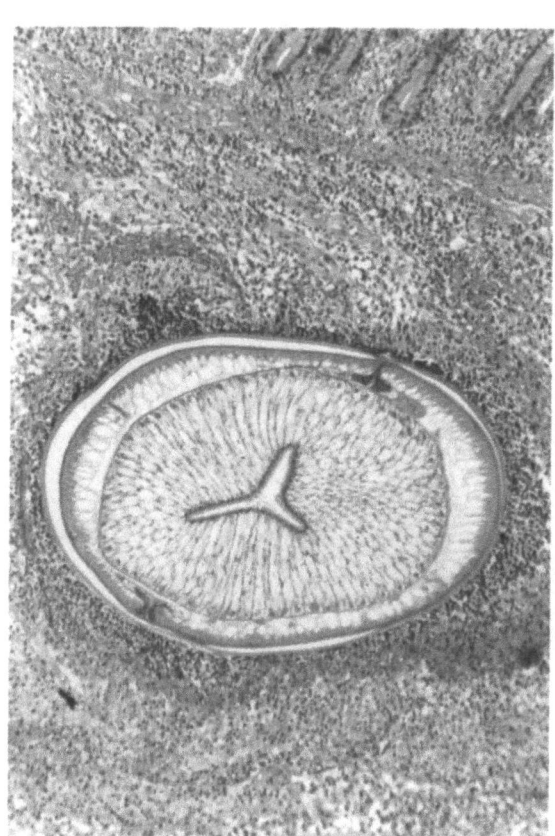

Figure 3.17. Cross section of an *Anisakis* larva invading the submucosa of the ileum. Note the characteristic lateral cords.

tions: anaphylactoid, Arthus type, and cell-mediated hypersensitivity. Advances in monoclonal antibody studies and immunohistology make it possible to detect the *Anisakis* specific antigen even in the decayed worm and thus to differentiate anisakidosis from other parasitic diseases.

Acute Intestinal Anisakidosis

The intestinal wall with acute intestinal anisakidosis is three to five times thicker than normal (Fig. 3.15), mainly due to intensive edema, especially in the submucosal and serosal layers; massive numbers of eosinophils, neutrophils, macrophages, and lymphocytes are seen. The cut surfaces of the larvae are found most frequently in the submucosal layer accompanied by the most intensive fulminant changes (Figs. 3.16, 3.17, 3.18). Even though *Anisakis* larva may not be found in one area of a phlegmonous lesion, they occasionally are found

Figure 3.18. Sagittal cut surface of an *Anisakis* larva invading into mucous membrane.

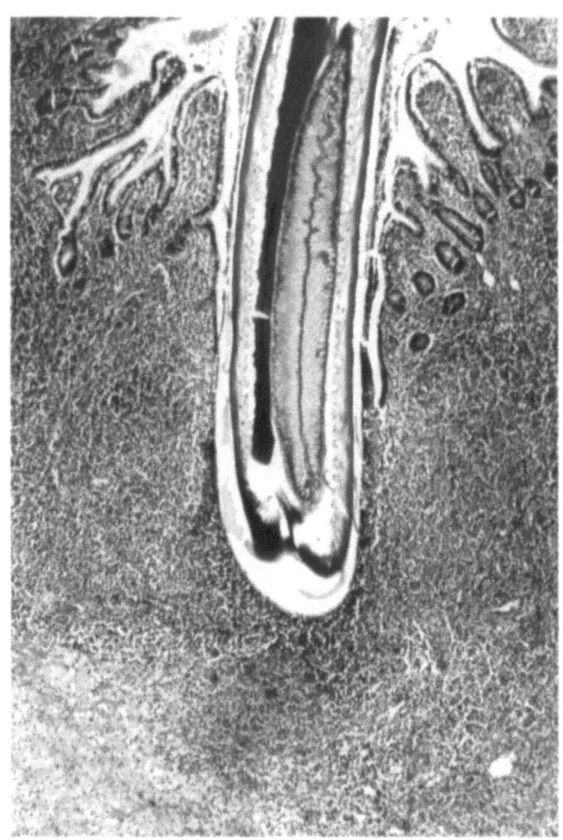

several centimeters nearby, and sometimes two or several cut surfaces can be seen because of the coil-shaped penetration of the larva or migration of more than one larva. The probability of discovering the larva depends on the number of tissue blocks examined. When the larva cannot be found by careful examination, the larva has probably escaped through the intestinal wall into the peritoneal cavity. The most intensive histologic changes observed around penetrating *Anisakis* larva are necrosis and marked cell infiltration including eosinophils. The larva is usually intact preserving its characteristic morphology, as with this type the acute reaction takes only about 48 hours after larval migration.

Histopathology of the lesion other than the site of larva penetration is, in general, characterized by intensive edema, marked eosinophil infiltration, fibrin exudation, small hemorrhages, and severe vasculitis with cell infiltration and fibrinoid necrosis, especially in the submucosal layer (Figs. 3.19 and 3.20). Ulceration or necrosis is not always detected in the mucosal layer. Hyperplasia of lymphoid fol-

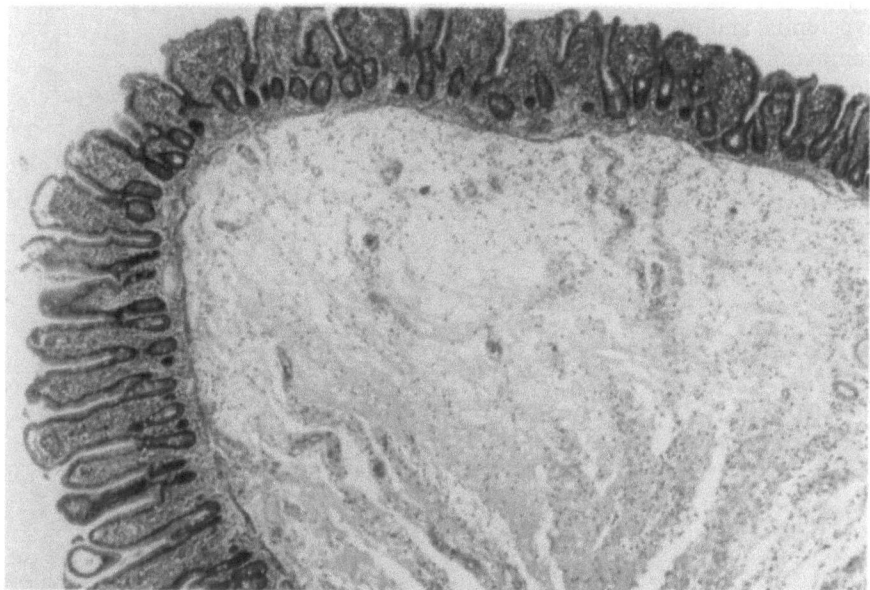

Figure 3.19. Marked edema of the submucosa of the ileum with intestinal anisakidosis.

licles is found in most cases. The muscle layer shows weaker inflammatory changes than the submucosal layer, without marked edema and fibrin exudation. Ganglion cells in the Auerbach plexus are often invaded by inflammatory cell infiltration, and they are swollen and degenerated (Fig. 3.21). The serosal layer may display massive cell infiltration, weak edema, and sometimes remarkable hemorrhage. In several cases the submucosal area is swollen with fibrosis, especially at the perivascular areas.

Chronic Intestinal Anisakidosis

Chronic intestinal anisakidosis is similar to a mild type of gastric anisakidosis, and its patholgic features are essentially the same as those in the stomach. This mild disease seems to be caused by prolongation of the allergic reaction due to long-standing larval antigen from the primary infection with bacterial contamination. Kojima's classification is also applied in this type of anisakidosis.

First stage—phlegmonous type: The histologic appearance of this type is almost indistinguishable from that of acute fulminant type of intesitinal anisakidosis.

Figure 3.20. Edema, hyperemia, and inflammatory cell infiltration in the submucosa.

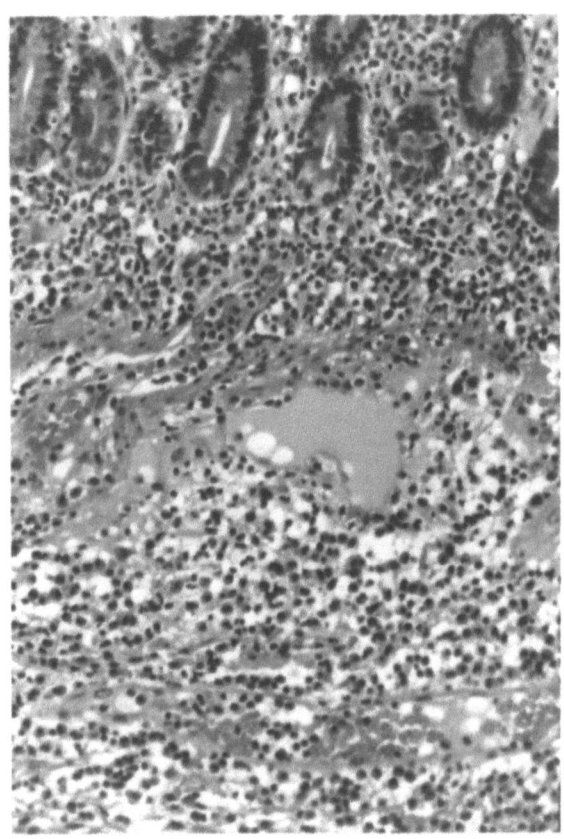

Second stage—abscess type: Most cases of this type are considered to arise from the Arthus-type allergy complicated by bacterial infection. Abundant eosinophils, neutrophils, macrophages, and lymphocytes surround the slightly degenerating larva or its debris (Fig. 3.22). The larva is surrounded by massive necrotic tissue in its inner layer, with infiltration of eosinophils into the larva. A small amount of granulomatous tissue around the abscess is infiltrated by inflammatory cells accompanied by edema and fibrin exudation.

Third stage—abscess-granuloma type: Most cases of this type are found incidentally when the patient with an intestinal disease (e.g., ileus, submucosal tumor, or chronic enteritis) has undergone surgery. When an abscess is long-standing, organization takes place to form granulomatous tissue. Intensively degenerated worm, which may not be identified as *Anisakis* larva, is observed at the center of this abscess-granuloma tissue (Fig. 3.23). Only an immunohistologic techinque with monoclonal antibody specific to *Anisakis* larva can

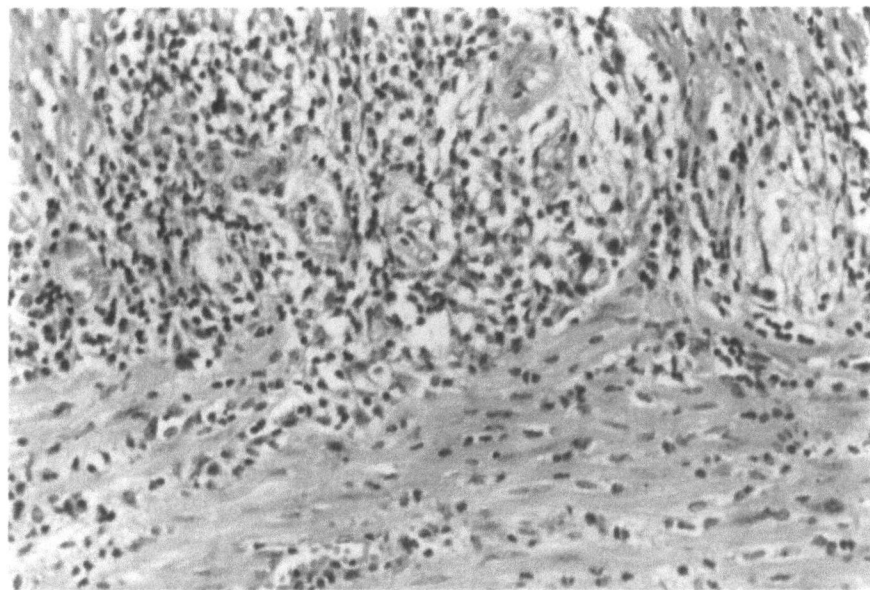

Figure 3.21. Ganglion cells in the Auerbach plexus are infiltrated by inflammatory cells that are swollen and degenerated.

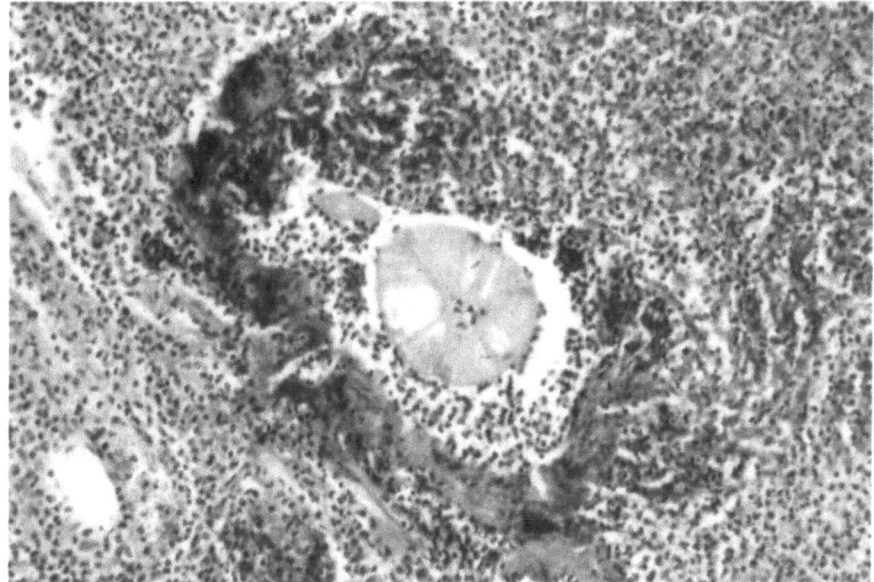

Figure 3.22. Decayed larva in the center of an abscess.

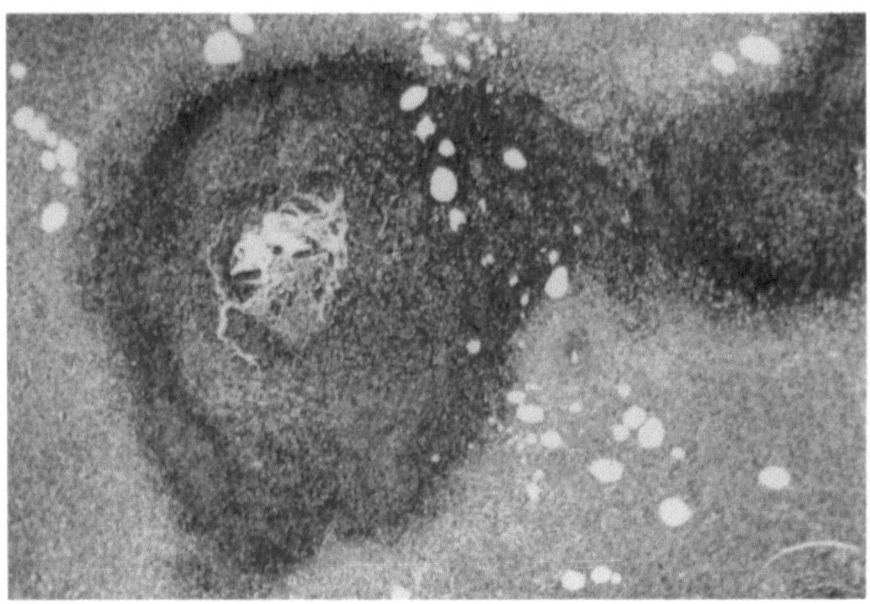

Figure 3.23. Intensively degenerated larva in abscess-granuloma tissue.

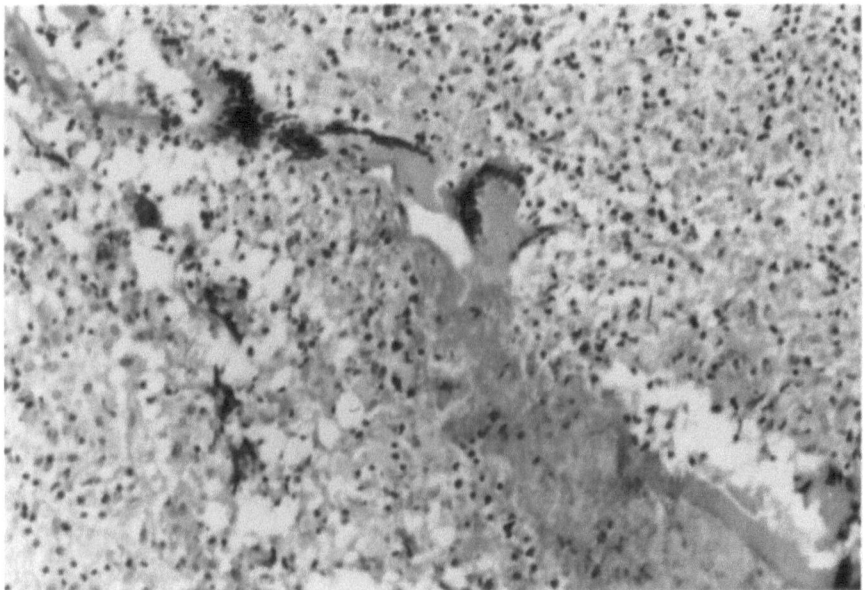

Figure 3.24. Unidentified material is invaded by eosinophils, neutrophils, and foreign body giant cells.

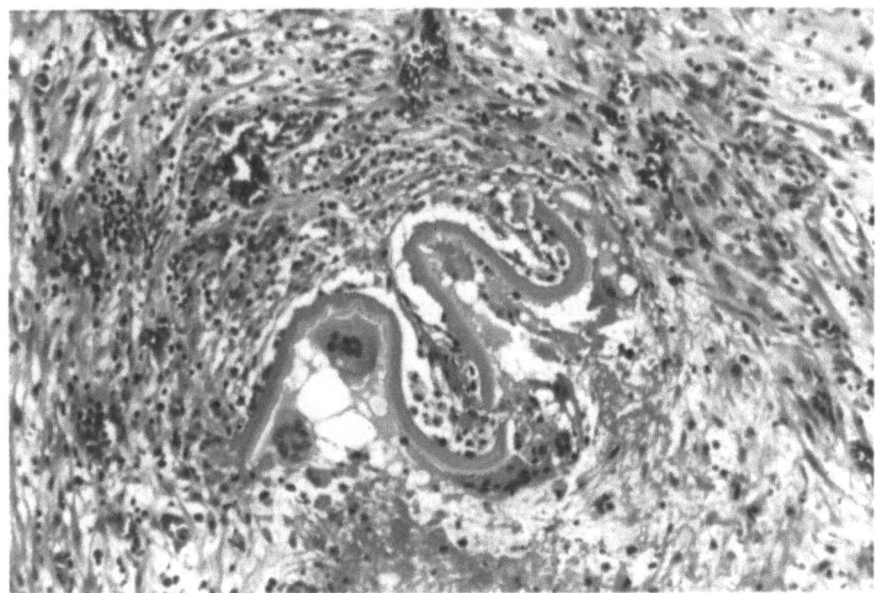

Figure 3.25. Intraabdominal *Anisakis* granuloma consisting of the *Anisakis* cuticle, foreign body giant cells, lymphoid cells, and fibrosis.

identify *Anisakis* antigen. This material is invaded by eosinophils and neutrophils and sometimes by epithelioid cells and foreign body giant cells (Fig. 3.24). Lymphocyte infiltration predominates.

Fourth stage—granuloma type: Unlike the corresponding type in the stomach, this type in the intestine is rarely encountered, as almost all patients with chronic intestinal anisakidosis are operated surgically and the larva resected before the disease reaches this endstage. Even if intestinal granuloma formation is found histopathologically, it does not lead to the diagnosis of anisakidosis, because detection of *Anisakis* larva and identification of the characteristic histopathologic features of anisakidosis are almost impossible. If some debris or unidentified materials are found in the granuloma, an immunohistologic technique with monoclonal antibodies specific to *Anisakis* larva can be applied. Serologic examination with monoclonal antibodies helps to establish a definite diagnosis.

Destruction of larva might progress systematically. The grade of larval destruction can be estimated by the duration of the disease. The pathologic stages of anisakidosis are not strictly separated but are continuous events, making such estimation difficult.

Larvae that have migrated into the intestinal wall may also disappear into the abdominal cavity or intestinal lumen, but the excretory-secretory (ES) antigen left in the intestinal wall can cause a secondary allergic reaction. In such a case, inflammatory changes take place rapidly without granuloma formation; intestinal resection is not performed. Sometimes extragastrointestinal *Anisakis* granulomas grow and are resected as an unidentified intraabdominal tumor of the omentum or mesentery. Histologic examination of these specimens reveals granuloma formation with decayed larva or with *Anisakis* cuticle sloughed from the larva in it (Fig. 3.25).

Immunology

Anisakidosis is a disease commonly found among people in Japan who consume raw fish (sushi or sashimi). Sushi is also becoming popular in the United States and in some European countries, and anisakidosis is being found in these places also. The etiologic organism, *Anisakis simplex* larvae, causes epigastric pain clinically and induces a striking eosinophilic infiltration histologically (99). However, the nature of the etiologic mechanism by which clinically manifested anisakidosis develops is still controversial. The diagnosis of anisakidosis can be made from the history of eating raw fish, and a final diagnosis is not easy even with the development of gastrofiberscopical examination of patients. This disease also becomes more complicated when it continues as a chronic form (100). Furthermore, it is more difficult to see the whole length of the intestine during endoscopy. Therefore there is an increasing demand for the development of a specific serodiagnostic assay system for this disease.

Monoclonal Antibodies Against Anisakis simplex Larvae

To establish the serodiagnostic assay system for anisakidosis, we first attempted to develop monoclonal antibodies (mAbs) that discriminate the antigen distribution specifically found in *Anisakis simplex* larvae. Fortunately, we obtained several mAbs that react specifically with the antigen of this specific parasite (101). Second, we utilized these mAbs for the development of the specific serodiagnosis for the anisakidosis using patients' sera.

The mAbs were developed by hybridizing hypoxanthine-aminopterine-thymidine (HAT)-sensitive mouse myeloma cell NS-1 with spleen cells of mice immunized with *Anisakis simplex* larvae.

Table 3.11 Reactivity of monoclonal antibodies against various parasites and human tissues.

	mAb reactivity						
Parasite	An1	An2	An3	An4	An5	An6	An7
Anisakis simplex larvae	+	+	+	+	+	+	+
Ascaris suum	−	−	−	−	−	±	−
Pseudoterranova decipiens	−	−	−	−	−	−	−
Toxocara cati	−	−	−	−	−	−	−
Trichinella spiralis	−	−	−	−	−	−	−
Echinococcus multilocularis	−	−	−	−	−	−	−
Dirofilaria immitus	−	−	−	−	−	−	−
Human tissues	−	−	−	−	−	−	−
Isotype	IgG1	IgG1	IgM	IgG1	IgM	IgG1	IgM

Table 3.12 Patterns of reactivities with monoclonal antibodies on frozen sections of *Anisakis* larvae.

		mAb reactivity				
mAb	Mol. (kDa) wt.	Muscle	Pseudocoel	Renette cells	Intestines Membrane	Intestines Cytoplasm
An1	34	+	−	+	−	−
An2	40, 42–46	+	±	+	+	−
An3	ND	+	±	±	−	−
An4	130	+	+	+	−	−
An5	ND	+	±	−	−	−
An6	68	+	−	±	−	+
An7	ND	−	−	+	+	−

ND = not determined.

We obtained seven mAbs, namely An1 through An7 (101) (Table 3.11). These mAbs react specifically with *Anisakis simplex* larvae, except An6, which also reacts weakly with *Ascaris suum*. These mAbs do not react with *Pseudoterranova decipiens* larvae, *Toxocara cati*, *Trichinella spiralis*, *Echinococcus multilocularis*, or *Dirofilaria immitus*. Western blot analysis of these mAbs-defined antigens was performed. *Anisakis simplex* larvae was homogenized in PBS, and the supernatant was centrifuged at $100,000 \times g$ for 1 hour and was used as *Anisakis* antigens. These antigens were run on sodium dodecyl sulfate polyacrylamide gel electrophoresis (SDS-PAGE) and electrically blotted to the membrane. We also estimated the molecular weight by gel filtration with Sephadex G-200. The data (Table 3.12) suggested that An1 mAb recognizes a single polypeptide chain with the molecular weight of 34 kilodaltone (kDa), as the electrophoretic mobility pattern was the same with or without the addition of 2-mercaptoethanol. An2 seemed to detect a disulfide-linked heterodimer composed of 40 and 42–46 kDa. The data also suggested that An4 and An6 recognized antigen molecules with 130 and 68 kDa molecular weight, respectively. We could not determine the molecular weight of other mAbs-defined antigens by these experimental protocol, suggesting that the antigens defined by An1, An2, An4, and An6 are water-soluble, whereas those defined by An3, An5, and An7 are not.

We analyzed the localization of antigens detected by these mAbs in frozen sections of *Anisakis simplex* larvae using fluorescein isothiocyanate (FITC)-conjugated anti-mouse immunoglobulins. The data indicated that these mAbs could recognize different antigenic epitopes present on *Anisakis simplex* larvae (Table 3.12). All mAbs except An7 reacted with muscle of *Anisakis simplex* larvae. An2 through An5 reacted also with pseudocoel. An1, An2, An3, and An7 strongly reacted with Renette cells. Furthermore, An2 could detect the antigen distributed in the membrane of the intestines. These data indicate that mAbs could be a powerful tool to examine the immunobiologic development of anisakidosis and the roles of each specific antigen detected by each mAb.

It has been suggested that ES antigen plays an important role in the pathologic development of anisakidosis. On frozen sections of *Anisakis simplex* larvae, An1, An2, and An6 seem to react with ES antigens. Among these three mAbs, An2 could react with this antigen when produced and secreted into the culture supernatant within 48

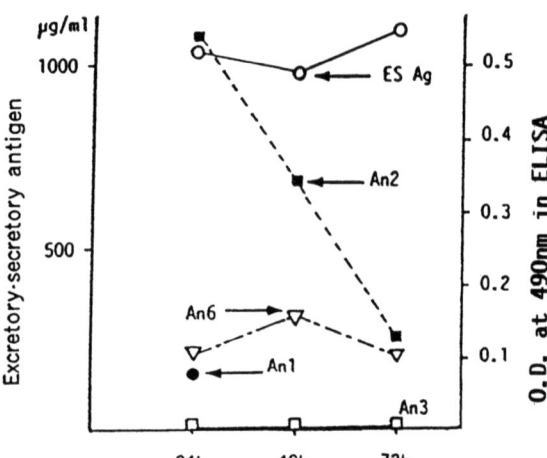

Figure 3.26. Reactivities of mAbs against ES antigen in a micro-ELISA. Protein content in ES antigen (open circles) was determined by the method of Lowry et al.

hours of larvae culture (Fig. 3.26). ES antigens appear to be rapidly denatured, as they have potent protease activity, indicating that the determinant detected by An2 has also disappeared. The antigenic epitopes recognized by An1 and An6 seem to have been destroyed, and these two mAbs could not react with ES antigen in the culture supernatant of *Anisakis simplex* larvae. These data suggest that An2 is a powerful mAb that can be used to develop a specific serodiagnostic assay system for anisakidosis.

Serodiagnosis by micro-ELISA for Anisakidosis Using An2.

We attempted to establish a serodiagnostic assay system by using An2 mAb, as this mAb could detect the specific epitope which is present on *Anisakis simplex* larvae. Using the micro enzyme-linked immunosorbent assay (micro-ELISA) and 93 serum samples from patients with anisakidosis at various clinical stages and at various times after the onset of clinical anisakidosis (from day 1 through week 12), we tested whether An2 can be used effectively for establishing a micro-ELISA procedure for the serodiagnosis of anisakidosis (102). For the micro-ELISA, about 100 µl of PBS containing of An2 100 mg/ml was adsorbed at 37°C for 1 hour in wells of the micro-ELISA plates (Sumitomo Beikleight, Tokyo). The mAb was first purified using protein A Affi-Gel beads (Bio-Rad Laboratories) (103,104). The plates were washed with PBS: and 50 µl of the antigen containing 100 µg protein/ml that was obtained by centrifuging the homogenized *Anisakis* larvae in PBS at 3000 rpm was applied to each well for 1 hour at 37°C. Plates were then washed several times with PBS. Diluted sera from 93 patients and 45 controls was then applied at 37°C for 1 hour before washing. Peroxidase-conjugated goat anti-human IgG, IgA, and IgM or anti-human IgE serum was reacted at 37°C for 1 hour in each well and then washed. Finally, 100 µl of solution containing 400 µg of O-phenylenediamine dihydrochloride (OPD; Nakarai Chemical, Kyoto) in 1 ml of substrate buffer (0.05 M citric acid, pH 4.0) was added and reacted in the presence of 0.01% H_2O_2 at 37°C for 30 minutes. Each well was counted at OD 490nm. To establish the diagnostic criteria in the micro-ELISA system, we set appropriate indices for IgG, IgA, IgM, and IgE by calculating:

O.D. anisakidosis patient's serum/O.D. control sera

Of 93 serum samples from patients with anisakidosis, Nos. 39, 15, 11, 8, 8, 2, 3, and 7 were obtained on day 1 and at 1, 2, 3, 4, 5, 8, and

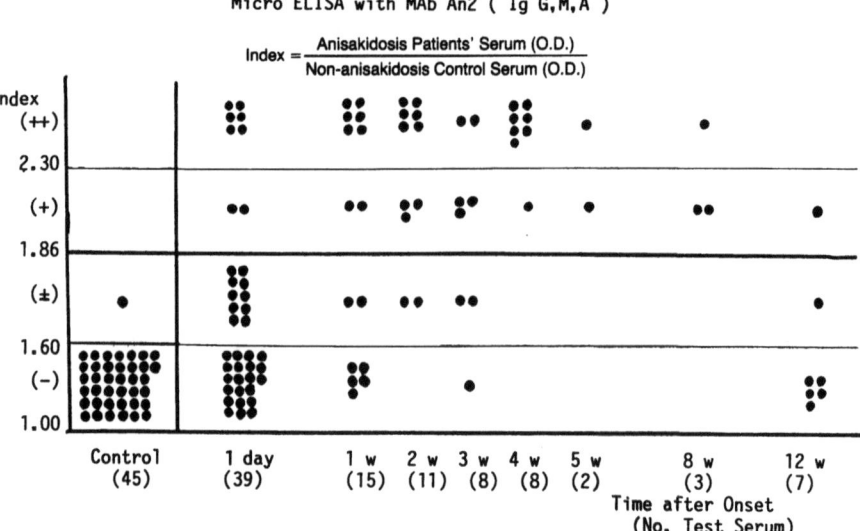

Figure 3.27. Reactivity in micro-ELISA of serum IgG, IgA, and IgM to the antigen immobilized by An2. Data shown are various phases in patients with anisakidosis and controls. Index = anisakidosis patients' serum (O.D.)/ nonanisakidosis (control) serum (O.D.) The numbers at the bottom of the graph represent the time after onset. The numbers in parentheses represent the number of test sera.

12 weeks, respectively, after onset of disease. As shown in Figure 3.27, the data clearly indicated that most patient IgG, IgA, and IgM reacted strongly with the An2 antigen at 1, 2, 3, 4, 5, and 8 weeks after onset of anisakidosis. In contrast, most, but not all, control serum samples showed weak reactivity in the assay system. Even at 1 day, many samples clearly reacted with the antigen. At 12 weeks after onset of disease, six of seven serum samples showed nearly the same range of reactivity as the control sera. When an index of 1.86 was given for serodiagnosis in the micro-ELISA, 100% diagnostic efficiency was obtained at 4, 5, and 8 weeks after onset of the disease. In contrast, all control sera except one sample were reasonably included in negative cases.

The reactivity of IgE from patients with anisakidosis and control are shown in Figure 3.28. The data showed that many patients reacted strongly with the An2 antigen as early as 1 day or 1 week after disease onset. Figure 3.28 also shows an index for the diagnostic criteria as IgE of sera detected by micro-ELISA. With an index of 1.72, all control sera were clearly categorized as negative cases. In contrast, many patients's sera clearly showed a high index value. These data indicate

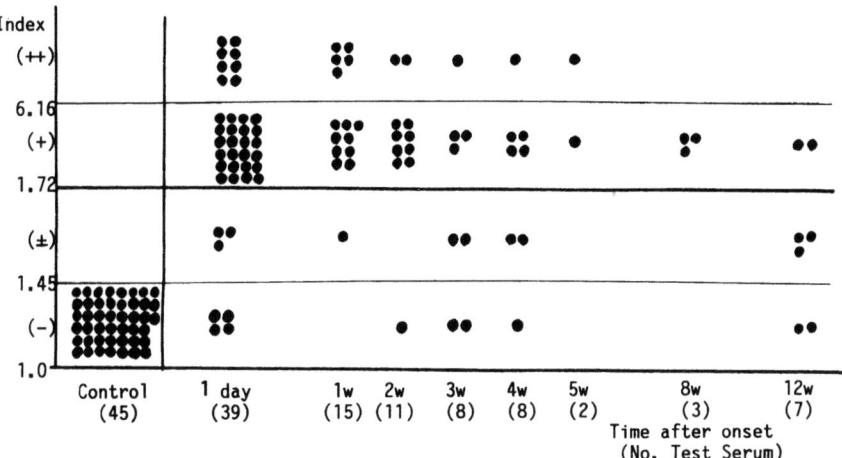

Figure 3.28. Reactivity by micro-ELISA of serum IgE to the antigen immobilized by An2. See Figure 3.27 of explanation of graph.

that serodiagnostic assays using An2 can provide the high diagnostic efficiency for human anisakidosis.

Significance of Serodiagnosis Using An2

The diagnosis of anisakidosis is not easy for physicians. Particularly, the diagnosis of the intestinal anisakidosis is almost impossible by usual endoscopic examinations. Therefore the micro-ELISA discussed here seems practical and important from a clinical point of view. Fortunately, we obtained several mAbs that could react specifically with *Anisakis simplex* larvae antigens. It was suggested that An2 reacted strongly with ES antigen, which is considered to play an important role in the development of clinical anisakidosis. This mAb detects a heterodimer composed of 40 and 42–46 kDa antigen molecules. It was interesting that patient's sera with severe clinical manifestation also detected the antigens with the same molecular size as An2 (data not shown), indicating that these patients' sera may react directly with An2-defined antigen. These data suggest that An2-defined antigen detected in this micro-ELISA is significant for detecting anisakidosis.

The titration of patients' IgE reacting with *Anisakis simplex* larvae antigen could be beneficial for diagnosing this disease. The titration of patient's IgG, IgA, and IgM may also contribute to the clinical diagnosis. The background activity to the antigen was relatively high in some control sera. Although these sera might be false-negative samples, we should reduce the background reactivity by technical improvements. Our preliminary data using a large number of test samples of patients' sera suggest that the current micro-ELISA system may be used to arrive at a prompt diagnosis of anisakidosis. We are now suggesting that this system be used as a routine laboratory assessment.

Molecular genetics of *Anisakis* larvae

It is well known that gastric and intestinal anisakidosis occurs when one ingests raw marine fishes, such as mackerels and codfishes, that are infected with third-stage larvae. Although the relation between adult worms of genus *Anisakis* and their larvae has been morphologically studied, some are still uncertain (105). It is now apparent that the techniques of molecular biology are useful for the identification of parasites and can provide a new approach to taxonomic study. Analysis of restriction fragment length polymorphisms (RFLPs) is an excellent method in distinguishing these larvae and for clarifying the relation between larvae and their adult worms, because it can be used to detect minute differences in genomic DNA sequences among closely related organisms (106). We studied the relation between *A. simplex* and *Anisakis* larvae type 1 by the analysis of RFLPs. We also compared RLFPs between *Anisakis* larvae type 2 and adult worms of *Anisakis physeteris*.

Collection of Worms

Anisakis larvae type 1 were recovered from the viscera of a paratenic host, codfishes, obtained from fish shops. Larvae were wahed four or five times with 0.85% saline, immediately frozen in liquid nitrogen, and stored at −85°C. Larval type was identified under the stereomicroscope according to the criteria of Koyama et al. (26). *Anisakis* larvae type 2 were collected from the viscera of bonitos. Adult worms of *A. simplex* were collected from the stomach of *Globicephala macrorhyncus*. Adult worms of *A. physeteris*, which had been collected from *Physeler macrocephalus*, were provided from Dr. Lia Paggi (Italy).

Preparation of Genomic DNA

Genomic DNAs were isolated from *Anisakis* larvae type 1, *Anisakis* larvae type 2, adult worms of *A. simplex*, and adult worms of *A. physeteris*. Each 100 mg of frozen larvae and adult worms in 1 ml of NET solution (0.1 M NaCl, 0.05 MEDTA, 0.1 M Tris HCl, 1% SDS, pH 8.0) were disrupted in a Freezer/Mill pulverizer (Spex Industries) for 10 minutes. The disrupted material was rapidly transferred to a siliconized tube and incubated with occasional gentle agitation at 65°C for 60 minutes in proteinase K solution at 1 mg/ml. After centrifugation at 10,000 rpm for 10 minutes at 0°C, the supernatant was collected in a sterile siliconized tube, and 4.5 g calsium choloride was dissolved in 5 ml of the supernatant by gentle mixing. Then ethidium bromide was added to the mixture at a final concentration of 600 μg/ml and centrifuged at 42,000 rpm for 36 hours at 20°C in a Backman 50 Ti rotor. A single DNA band was clearly visualized under ultraviolet illumination. The DNA band was recovered, and the ethidium bromide in the collected DNA solution was extracted with an equal volume of isoamyl alcohol. After dialysis against TE buffer (10 mM Tris HCl, 1 mM EDTA, pH 7.5) overnight, the DNA in the aqueous phase was precipitated by two volumes of cold ethanol in the presence of 0.15 M NaCl overnight at -80°C, dried in a vacuum, and dissolved in a small volume of TE buffer.

Preparation of 32p-25s rRNA

Total RNA was isolated from *Anisakis* larvae type I from codfishes by the guanidinium/hot phenol method. Samples containing rRNA were electrophoresed through 1.0% agarose gel containing formaldehyde at 200 V for 2 hours in MOPS buffer, as described by Maniatis et al. (107). Two bands of 25s and 17.5s rRNA can be clearly visualized after staining the gel with ethidium bromide (1 μg/ml) (106). The 25s rRNA was recoverd from the gel as described by McDonnel et al. (108). The T4 polynucleotide kinase method described by Maxman and Gilbert (109) was used for labeling 25s rRNA with γ^{32}P-ATP (Amersham International).

Digestion with Restriction Endonuclease and Electrophoresis

Genomic DNA (0.1 μg) of larvae or adult worms was digested with EcoRI, HaeIII, or HpaII, according to the manual of the supplier of these enzymes (Toyobo Inc., Tokyo). Restriction fragments after EcoRI digestion were electrophoretically separated on 0.8% agarose

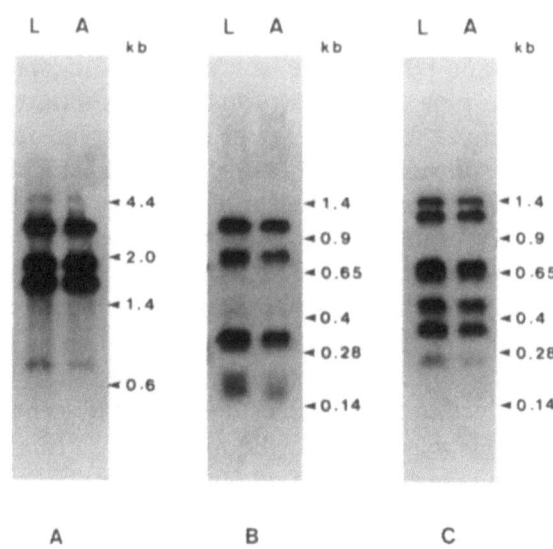

Figure 3.29. Restriction fragment length polymorphisms of *Anisakis* larvae type I and adult worms of *Anisakis simplex*. Genomic DNA was digested with EcoRI (A), Hae III (B), and Hpa II (C) and was analyzed as described in Materials and Methods. (Lane L) *Anisakis* larvae type I from codfish. (Lane A) Adult worms of *Anisakis simplex*. Arrows indicate position of markers.

gel, and those after HaeIII or HpaII digestion were separated on 1.5% agarose gel.

Southern blot analysis

Restriction fragments separated by agarose gel electrohoresis were transferred to a nylon membrane (Hybond-N; Amersham International) using the method described by Southern with some modification and hybridized with ^{32}P-25s rRNA probe according to Amersham's manual.

RFLPs of *Anisakis* Larvae Type I and Adult Worms of *A. simplex*

According to morphologic studies, *Anisakis* larva type I is considered to be the larva of *A. simplex* (36). As shown in Figure 3.29, the patterns of these two kinds of DNA fragment were exactly the same in three different endonuclease digestions, respectively.

RFLPs of *Anisakis* Larvae Type II and Adult Worms of *A. physeteris*

We demonstrated that *Anisakis* larvae type II and adult worms of *A. physeteris* have the same RFLP patterns (Fig. 3.30).

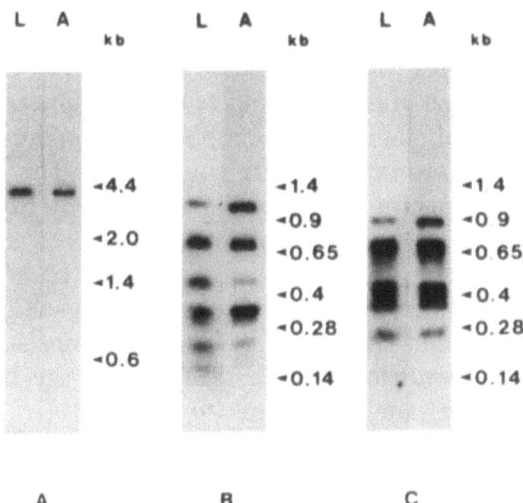

Figure 3.30. Restriction fragment length polymorphisms of *Anisakis* larvae type II and adult worms of *Anisakis physeteris*. Genomic DNA was digested with EcoRI (A), Hae III (B), and Hpa II (C) and analyzed as described in Materials and Methods. (Lane L) *Anisakis* larvae type II from bonito. (Lane A) Adult worms of *Anisakis physeteris*. Arrows indicate position of markers.

Genetic Discrimination Between Adult Worms and Larvae

In our previous study there were no differences in the RFLPs of two kinds of *Anisakis* larvae type I genomic DNA from different paratenic hosts, and we demonstrated the same result in *Anisakis* larva type II (data not shown). Those results suggested that genomic DNAs of *Anisakis* larvae were not affected by environmental conditions. Moreover, the clear differences among the *Anisakis* larvae types I and II and *Contracaecum* larvae in terms of the RFLPs of genomic DNAs were shown by Southern blot assay (106).

In this paper, we studied the relation between adult worms and their larvae. RFLPs of *Anisakis* larvae type I and adult worms of *A. simplex* were the same. Also, the same RFLP patterns were demonstrated for *Anisakis* larvae type II and adult worms of *A. physeteris*. These results suggest that *Anisakis* larva type I must be the larva of *A. simplex* and type II must be that of *A. physeteris*. These results provide agreement with morphologic studies and indicate that 25s rRNA will prove useful in similar studies.

Conclusion

We have introduced on this review the taxonomy, morphology, and life cycles of *Anisakia*. In the epidemiology section, we described anisakidosis cases reported worldwide to date and discussed the reason there are more cases of gastric anisakidosis than intestinal anisakidosis in Japan. In addition, the changing aspects of the species of intermediate host and infection rates were discussed.

The symptomatology of human anisakidosis have already appeared in reviews from various countries. This review emphasized some unique clinical manifestations sometimes seen with this disease. The relations of anisakidosis with vanishing tumor of the stomach, AGML, and intraperitoneal and ectopic anisakidosis was discussed.

We also described the pathology of human anisakidosis. Human anisakidosis has been classified into gastric and intestinal anisakidosis, each having acute and chronic forms. We described diagnosis by the ELISA method utilizing monoclonal antibodies and by the immunofluorescent method using tissue specimens.

During the last few years, studies on anisakidosis and anisakidae, enthusiastically revitalized in Western countries, were presented at workshops at Kiel University in 1989 and in Canada in 1990. Studies on Anisakidae siblings using the genetic approach have also been published (35). Two volumes on gastric and intestinal anisakidosis in Japan edited by us (70,71) have been published, with an update of the academic status of the study of anisakidosis. The recent progress in the study of this disease is so remarkable that there may be much hope for better treatment in the future.

References

1. Ishikura H, Asanuma T: On the strange terminal ileitis accompanied with acute ileocaecal syndromes (first report). J Hokkaido Branch Jpn Surg Soc 1958;3:63–64 [in Japanese].
2. Ishikura H, Tanaka M: On the strange terminal ileitis acompanied with acute ileocaecal syndromes (second report). J Hokkaido Branch Jpn Surg Soc 1959;5:37–38 [in Japanese].
3. Otsuru M, Ishizuki F, Hatsukano T: Regional ileitis caused by invasion of Ascalis larva into intestinal wall. Nippon Ijishinpo 1957;1775:25–38 [in Japanese].
4. Beaver PC: Chronic eosinophilia due to visceral larva migrants. Pediatrics 1952;9:7–19.
5. Van Thiel PH, Kuiper FC, Roskam RT: A nematode parasitic to herring causing acute abdominal syndromes in man. Trop Geogr Med 1960; 12:97–113.
6. Van Thiel PH: Anisakiasis. Parasitology 1962;52(suppl):16–17.
7. Namiki M, Morooka T, Kawauchi H, et al: Diagnosis of acute gastric anisakiasis. Stomach and Intestine 1970;5:1437–1440.
8. Ishikura H: Current topics on Anisakiosis—popularization, decide diagnostic, and specific antigen. Jpn J Gastroenterol 1990;87:1740–1743 [in Japanese].

9. Desowitz RS: Human and experimental anisakiasis in the United States. Hokkaido J Med Sci 1986;61:358–371.
10. Ishikura H: Clinical and immunological studies on anisakiasis. Hokkaido J Med Sci 1968;43:83–99 [in Japanese with English abstract].
11. Iwano H, Ishikura H, Hayasaka H: Statistical analysis of anisakiasis in Japan during the last five years. Geka Shinryo 1974;16:1336–1342 [in Japanese].
12. Hirata F, Adachi H, Ito Y, et al: Investigation of acute gastric change caused by *Anisakis* larva. Gastroenterol Endosc 1984;26:2135–2136.
13. Omachi K, Omachi T, Maruyama Y: Anisakiasis of gastrointestinal tract in Nagano Prefecture. J Shinsyu Med Soc 1985;33:42–56 [in Japanese with English abstract].
14. Hartwich G: Key to genera of the Ascaridoidea. In Anderson C, Chabaud AG, Willmott S (eds). CH1 Key to the Nematode Parasites of Vertebrates. Common-wealth Institute of Helminthology, Headly Brothers Ltd., London, 1974, No.2, pp. 2–5.
15. Davey JT: A revision of the genus Anisakis Dujardin, 1845 (Nematoda: Ascaridata). J Helminthol 1971;45:51–72.
16. Nascetti G, Paggi L, Orecchia P, et al: Electrophoretic studies on the Anisakis simplex complex (ASCARIDIA: ANISAKIDAE) from the Mediteranean and North-East Atrantic. Int J Parasitol 1986;16:633–640.
17. Nascetti G, Paggi L, Orecchia P, et al: Divergenza genetica in popolazioni del genere Anisakis del Mediterraneo. Parassitologia 1981;23:208–210.
18. Nascetti G, Paggi L, Orecchia P, et al: Two sibling species within Anisakis simplex (Ascaridida: Anisakidae). Parassitologia 1983;25:306–307.
19. Orrecchia P, Paggi L, Mattiucci S, et al: Electro phoretic identification of larvae and adults of Anisakis (Ascaridida: Anisakidae) J Helminthol 1986;60:331–339.
20. Mattiucci S, Nascetti G, Bullini L, et al: Genetic structure of Anisakis physeteris, and its differentiation from the Anisakis simplex complex (Ascaridida: Anisakidae). Parasitology 1986;93:383–387.
21. Mosgovoy AA: Ascarids of mammals of the U.S.S.R. Tr Gel'minthol Lab. 1951;5:12–22 [in Russian].
22. Gipson DI: The systematics of ascaridoid nematodes: a current assessment. In Stone AF, Platt HM, Khalil LF(eds): Nematode Systematics Association. Special Volume No.22. Academic Press, London, 1983, pp. 321–338.
23. Gibson DI, Colin JA: The Terranova enigma. Parasitology 1981;85(Proc. B.S.T.):xxxxvi–xxxvii.
24. Shiraki T: Larval nematodes of family anisakidae (NEMATODA) in the northern Sea of Japan as a causative agent of eosinophilic phlegmone or granuloma in the human gastrointestinal tract. Acta Med Biol 1974;22:57–98.
25. Berland B: Nematodes from some Norwegian marine fishes. Sarsia 1961;2:1–50.
26. Koyama T, Kobayashi A, Kumada M, et al: Morphological and taxo-

nomical studies on Anisakidae larvae found in marine fishes and squids. Jpn J Parasitol 1969;18:466–487 [in Japanese].
27. Grabda J: Studies on the cycle and morphogenesis of Anisakis simplex (Rudolphi, 1809) (Nematoda: Anisakidae) culture in vitro. Acta Ichtyol Piscat 1976;6:119–141.
28. Beverly-burton M, Nyman OL, Pippy JHC: The morphology, and some observations on the population genetics of Anisakis simplex larvae (Nematoda: Ascaridata) from fishes of the North Atlantic. J Fish Res Board Can 1977;34:105–112.
29. Banning PV: Some notes on a successful rearing of the herring-worm Anisakis marina L. (Nematoda: Heterocheilidae). J Cons Int Explor Mer 1971;34:84–88.
30. Pippy JHC, Banning PV: Identification of Anisakis larvae (1) as Anisakis simplex (Rudolphi, 1809, det. Krabbe 1878) (Nematoda: Ascaridata). J Fish Res Board Can 1975;32:29–32.
31. Carvajal J, Barros C, Santander G, Alcalde C: In vitro culture of larval anisakid parasites of the Chilean hake Merluccius gayi. J Parasitol 1981;67:958–959.
32. Oshima T, Oya S, Wakai R: In vitro cultivation of Anisakis type I and type II larvae collected from fishes caught in Japanese coastal waters and their identification. Jpn J Parasitol 1982;31:131–134.
33. Suzuki T, Ishida K: Anisakis simplex and Anisakis physeteris: physicochemical properties of larval and adult hemoglobins. Exp Parasitol 1979;48:225–234.
34. Sakaguchi Y, Katamine D: Survey of anisakid larvae in marine fishes caught from the East China Sea and the South China Sea. Trop Med 1971;13:159–169 [in Japanese].
35. Paggi L, Orecchia P, Bullini L, et al: Electrophoretic identification on Anisakis larvae from Mediterranean and North Atrantic. Parasitologia 1983;25:315–316.
36. Oshima T: Anisakis and anisakiasis in Japan and adjacent area. In Morishita K, Komiya Y, Matsubayashi H (eds): Progress of Medical Parasitology in Japan. Vol.4. Meguro Parasitological Museum, Tokyo, 1982, pp. 301–393.
37. Shimazu T: 1. Larvae of Anisakinae. 2. Ecology. In Jpn. Soc. Sci. Fish: Fish and Anisakis. Fish Scienes Series 7. Koseisha Koseikaku, Tokyo, 1974, pp. 23–43 [in Japanese].
38. Smith JW: Anisakis simplex (Rudolphi, 1809, det. krabbe, 1878) (Nematoda: Ascaridoidea): morphology and morphometry of larvae from euphausiids and fish, and a review of the life-history and ecology. J Helminthol 1983;57:205–224.
39. Nagasawa K: The life cycle of Anisakis simplex: a review. In Ishikura H, Kikuchi K (eds): Intestinal Anisakis in Japan. Springer-Verlag, Tokyo, 1974, pp. 31–40.
40. McClelland G, Misra RK, Marcogliese DJ: Variations in abundance of larval anisakines, sealworm (Phocanema decipiens) and related species in cod and flatfish from the southern Gulf of St. Lawrence (4T) and

the Breton Shelf (4Vn). Can Tech Rep Fish Aquat Sci 1983; No. 1021: x+51 pp.
41. Hafsteinsson H, Rizvi SSH: A review of the sealworm problem: biology, implications and solutions. J Food Prot 1987;50:70–84.
42. Uspenskaya AV: Parasite fauna of benthic crustaceans from the Barents Sea. Izdatel'stvo Akademiya Nauk SSSR, Moscow, 1963, pp. 127 [in Russian].
43. Shiraki T, Hasegawa H, Kenmotsu M, Otsuru M: Larval anisakid nematodes from prawns, Pandalus spp. Jpn J Parasitol 1976;25:148–152.
44. Smith JW: Thysanoessa inermis and T. longicaudata (Euphausiidae) as first intermediate hosts of Anisakis sp. (Nematoda: Ascaridata) in the northern North Sea, to the north of Scotland and at Faroe. Nature 1971;234:478.
45. Smith JW: Larval Anisakis simplex (Rudolphi, 1809, det. Krabbe, 1878) and larval Hysterothylacium sp. (Nematoda: Ascaridoidea) in euphausiids (Crustacea: Malacostraca) in the North-East Atlantic and northern North Sea. J Helminthol 1983;57:167–177.
46. Sluiters JF: Anisakis sp. larvae in the stomach of herring (Clupea harengus L.). Z Parasitenk 1974;44:279–288.
47. Oshima T, Shimazu T, Koyama H, Akabane H: On the larvae of the genus Anisakis (Nematoda: Anisakidae) from the euphausiids. Jpn J Parasitol 1969;18:241–248 [in Japanese with English summary].
48. Shimazu T, Oshima T: Some larval nematodes from euphausiid crustaceans. In Takenouti V, et al (eds): Biological Oceanography of the Northern North Pasific Ocean Dedicated to Shigeru Motoda. Idemitsu Shoten, Tokyo, 1972, pp. 403–409.
49. Shimazu T: Some helminth parasites of marine planktonic invertebrates. J Nagano-ken Junior Coll 1982;37:11–29.
50. Kagei N: Studies on anisakid Nematoda (Anisakidae). IV. Survey of Anisakis larvae in the marine Crustacea. Bull Inst Publ Health 1974;23:65–71 [in Japanese with English summary].
51. Kagei N: Euphausiids and their parasites (1). Geiken Tsushin 1979;No.329:53–62 [in Japanese].
52. Hurst RJ: Marine invertebrate host of New Zealand Anisakidae (Nematoda). NZ J Mar Freshw Res 1984;18:187–196.
53. Oshima T: A study on the first intermediate hosts of Anisakis. Saishin lgaku 1969; 24:401–404 [in Japanese].
54. Kagei N: A list of fish infected with larval nematodes of the subfamily Anisakinae. In: Fish and Anisakis. Fish Science. Series 7. Koseisha Koseikaku, Tokyo, 1974, pp. 98–107. [in Japanese].
55. Dailey MD, Brownell RL Jr: A checklist of marine mammal parasites. In Ridgway SH (ed): Mammals of the Sea. Biology and Medicine. Charles C Thomas, Springfield, IL, 1972, pp. 528–589.
56. Kagei N, Kureha K: Studies on anisakid Nematoda (Anisaidae). Survey of Anisakis sp. in marine mammals collected in the Antarctic Ocean. Bull Inst Publ Health 1970;19:193–196 [in Japanese with English summary].

57. Kagei N, Asano K, Kihata M: On the examination against the parasites of Antarctic krill, Euphausia superba. Sci Rep Whales Res Inst 1978;30:311–313.
58. Hatsushika R: An experimental study on development and hatching of the eggs of Anisakis physeteris (Nematoda: Ascaridata). Kawasaki Med J 1979;5:1–7.
59. Meyers BJ: On the morphology and life history of Phocanema decipiens (Krabbe, 1878) Myers, 1959 (Nematoda: Anisakidae). Can J Zool 1960;38:331–334.
60. McClelland G: Phocanema decipiens (Nematoda: Anisakidae): experimental infections in marine copepods. Can J Zool 1982;60:502–509.
61. Valter ED, Popova TJ: Role of the polychaete Lepidonotus squamatus in the biology of anisakids. Biol White Sea Moscow 1974;4:177–182 [in Russian].
62. Valter ED: An occurrence of Terranova decipiens (Nematoda, Ascaridata) in the amphipod Caprella septentrionalis Kroeyer. Moscow Univ Biol Sci Bull 1978;33:9–11.
63. Bjorge AJ: An isopod as intermediate host of cod-worm. FiskDir Skr Ser HavUnders 1979;16:561–565.
64. Scott DM: Experimental infection of Atlantic cod with a larval marine nematode from smelt. J Fish Res Board Can 1954;11:894–900.
65. Burt MDB, Campbell JD, Likely CG: Serial passage of larval Pseudoterranova decipiens (Nematoda; Ascaridea) in fish. Can J Fish Aquat Sci 1990;47:693–695.
66. Dailey MD, Jensen LA, Hill BW: Larval anisakine roundworms of marine fishes from southern and central California, with comments on public health significance. Calif Fish Game 1981;67:240–245.
67. McClelland G: Phocanema decipiens: pathology in seals. Exp Parasitol 1980;49:405–419.
68. McClelland G: Phocanema decipiens: molting in seals. Exp Parasitol 1980;49:128–136.
69. McClelland G: Phocanema decipiens: growth, reproduction, and survival in seals. Exp Parasitol 1980;49:175–187.
70. Ishikura H, Namiki M (eds): Gastric Anisakiasis in Japan. Epidemiology, Diagnosis, Treatment, Springer-Verlag, Tokyo, 1989, pp. 1–144.
71. Ishikura H, Kikuchi K (eds): Intestinal Anisakiasis in Japan. Infected Fish, Seroimmunological Diagnosis and Prevention. Springer-Verlag, Tokyo, 1990, pp. 1–265.
72. Gefter WI, Boucot KR, Marshall EW: Localized interlobar effusion in congestive heart failure: vanishing tumor of the lung. Circulation 1950;2:336–343.
73. Kaye JI, Stassa G: Mimicry and deception in the diagnosis of the tumors of the gastric carcinoma. AJR 1970;110:295–303.
74. Wright FW, Matthews JM: Hemophilic pseudotumor of the stomach. Radiology 1971;98:547–549.
75. Yamazaki M, Hara K, Hasegawa T, Kanazawa M: Vanishing tumor of the stomach? J Clin Radiol 1976;21:47–54 [in Japanese].

76. Katz D, Siegel HL: Erosive gastritis and acute gastrointestinal mucosal lesion. In Glass GB (ed): Progress in Gastroentology. Vol. 1. Orlands, FL: Grune & Stratton, 1968, pp. 67-96.
77. Kawai K, Akasaka Y, Kimoto K, et al: Clinical aspects of acute upper-1 lesions, especially seen from the standpoint of hematemesis. Stomach Intestine 1973;8:17-23.
78. Yoshimura H, Akao N, Kondo K, Ohnishi V: Clinico-pathological studies on larval anisakiasis with special reference to the report of extra-gastrointestinal anisakiasis. Jpn J Parasitol 1979;28:347-354.
79. Yoshimura H: Clinical patho-parasitology of extra-gastrointestinal anisakiasis. Ishikura H, Kikuchi K (eds): Intestinal Anisakiasis in Japan. Springer-Verlag, Tokyo, 1990, pp. 146-154.
80. Ishikura H, Kikuchi Y, Hayasaka H: Pathological and clinical observation on intestinal anisakiasis. Arch Jpn Chir 1968;36:663-679 [in Japanese with English abstract].
81. Hayasaka H, Iwano H, Takagi R, et al: Studies on anisakiasis—on the influence of alcohol and isothiocyanate on the Anisakis larvae. Hokkaido J Surg 1969;14:167-171 [in Japanese with English abstract].
82. Yamada G: Studies on the prevention of Anisakis larva infection. J Osaka City Med 1971;20:131-150.
83. Kasuya S, Goto C, Koga K, et al: Lethal efficacy of components of ginger or gingerol and shogaol on Anisakis larva. Jpn J Parasitol 1989;38(suppl): 92 [in Japanese].
84. Kasuya S, Goto C, Koga K, et al: lethal efficacy of leaf extract from Perilla frutescens (traditional Chinese medicine) or perillaldehyde on Anisakis larvae in vitro. Jpn J Parasitol 1990;39:220-225.
85. Murata I, Yasuda I: Mortal or inhibitory effects with chinese medicine on the activity of Anisakis type 1 larvae. Jpn J Parasitol 1989;38(suppl): 43 [in Japanese].
86. Tsuji Y, Ishikura H, Kikuchi K, et al: Studies on regional Ileitis. 3. Pathological studies on regional ileitis in Hokkaido. Geka Chiryo [Surg Treat] 1965;13:144-154 [in Japanese].
87. Kojima K, Koyanagi T, Shiraki K: Pathological studies of anisakiasis (parasitic abscess formation in gastrointestinal tract). Jpn J Clin Med 1966;24:134-143 [in Japanese].
88. Smith JW, Wootten R: Anisakis and anisakiasis. In Lumsden WHR, Muller R, Baker JR (eds): Advances in Parasitology. Academic Press, London, 1978, pp. 93-163.
89. Sakanari JA, Loinaz HM, Deardorff TL, et al: Intestinal anisakiasis: a case diagnoses by morphologic and immunologic methods. Am J Clin Pathol 1988;90:102-113.
90. Kikuchi Y, Ishikura H, Kikuchi K, Hayasaka H: Pathology of gastric anisakiasis. In Ishikura H, Namiki M (eds): Gastric Anisakiasis in Japan. Epidemiology, Diagnosis, Treatment. Springer Verlag, Tokyo, 1989, pp. 117-127.
91. Kikuchi Y, Ishikura H, Kikuchi K: Pathology of intestinal anisakiasis. In Ishikura H, Kikuchi K (eds): Intestinal Anisakiasis in Japan. Infected

Fish, Seroimmunological Diagnosis, and Prevention. Springer Verlag, Tokyo, 1990, pp. 129–143.
92. Shibata O, Uchida Y, Furusawa T: Acute gastric anisakiasis with special analysis of the location of the worms penetrating the gastric mucosa. In Ishikura H, Namiki M (eds): Gastric Anisakiasis in Japan. Epidemiology, Diagnosis, Treatment. Springer Verlag, Tokyo, 1989, pp. 53–57.
93. Yazaki Y, Namiki M: Biopsy of gastric anisakiasis with acute symptoms. In Ishikura H, Namiki M (eds): Gastric Anisakiasis in Japan. Epidemiology, Diagnosis, Treatment. Springer Verlag, Tokyo, 1989, pp. 113–116.
94. Sasaki K, Sasaki T, Nagamine Y: A case report of intestinal anisakiasis with skip lesion and mesentric granuloma formation. J Jpn Soc Clin Surg 1984;45:1183–1187 [in Japanese].
95. Kikuchi Y, Ueda T, Yoshiki T, et al: Experimental immunopathological studies of intestinal anisakiasis. Igaku no Ayumi 1967;62:731–736 [in Japanese].
96. Suzuki T, Ishikura H: Studies on the etiologic mechanisms: symptoms and diagnosis of anisakiasis. Fishes Anisakis 1974;6:58–72 [in Japanese].
97. Shiraki T: On the pathological diagnosis of gastrointestinal larva migrans of anisakiasis. Saishin Igaku 1969;24:378–389.
98. Saeki H, Mizugaki H, Ishikura H, Hayasaka H: Immunological studies on anisakiasis. 2. Participation of immune response in host-tissue reaction and destruction of parasite bodies. Hokkaido J Med Sci 1972; 47:541–550 [in Japanese].
99. Tanaka J, Torisu M: Anisakis and eosinophil. I. Detection of a soluble factor selectively chemotactic for eosinophils in the extract from Anisakis larvae. J Immunol 1978;120:745–749.
100. Ashby BT, Appleton PJ, Dawson I: Eosinophilic granuloma of gastrointestinal tract caused by herring parasite Eustoma rotundatum. Br Med J 1964;1:1141–1145.
101. Takahashi S, Sato N, Ishikura H: Establishment of monoclonal antibodies that disseminate the antigen distribution specifically found in Anisakis larvae (type 1). J Parasitol 1986;72:960–962.
102. Yagihashi A, Sato N, Takahashi S, et al: A serodiagnostic assay by microenzyme-linked immunosorbent assay for human anisakiasis using a monoclonal antibody specific for Anisakis larvae antigen. J Infect Dis 1990;161:995–998.
103. Yagihashi A, Sato N, Torigoe T, et al: Identification of transformation-associated cell surface antigen expressed on the rat fetus-derived fibroblast. Cancer Res 1988;48:2798–2804.
104. Sato N, Sato T, Takahashi S, et al: Identification of transformation-related antigen by monoclonal antibody on Swiss 3T3 cells induced by transfection with murine cultured colon 36 tumor DNA. J Natl Cancer Inst 1987;78:307–313.
105. Ishikura H: General survey of Anisakis and anisakiasis. In Ishikura H, Namiki M (eds): Gastric Anisakiasis in Japan. Springer-Verlag, Tokyo, 1989, pp. 3–11.

106. Sugane K, Matsuura T: Restriction fragment length polymorphisms of Anisakinae larvae. J Helminthol 1989;63:269–274.
107. Maniatis T, Fritsch EF, Sambrook J: Molecular Cloning: A laboratory Manual. Cold Spring Harbor Laboratory, Cold Spring Harber, NY, 1982, pp. 194–203.
108. McDonnel MW, Simon MN, Studier FW: Analysis of restriction fragmemnts or T7 DNA and determination of molecular weights by electrophoresis in neutral and alkaline gels. J Mol Biol 1977;110:119–128.
109. Maxman AM, Gilbert W: Sequencing end-labeled DNA with base-specific chemical cleavages. Methods Enzymol 1980;65:499–506.
110. Tsukamoto H, Hiki Y, Mieno H, et al: Pathogenesis of acute gastric mucosal lesion. Stomach and Intestine 1989;24:645–651 (in Japanese with English abstract).
111. Saigenji K, Ozaki Y: Clinical study on AGML induced by upper G-I endoscopy-survery of questionaire in Japan. Gastroenterol Endosc 1989;31:785–790 (in Japanese with English abstract).
112. Matsushita F, Shibue T, Samejima Y, et al: Fifty two cases of acute gastric mucosal lesion observed after endoscopy. Gastroenterol Endosc 1986;28:717–727 (in Japanese with English abstract).
113. Iwashita A, Miyagahara T, Matsukuma T, et al: Histopathological study on acute gastric mucosal lesions after transcatheter arterial chemoembolization in patients with hepatocelluler carcinoma (in Japanese with English abstract).
114. Kitajima M, Kikuchi T, Sakai N, et al: Concepts on acute gastric mucosal lesions: Surgical viewpoints: Stomach and Intestine 1989;24:619–628 (in Japanese with English abstract).
115. Harada K: AGML: Its causes and clinical aspects. Stomach and Intestine 1989;24:637–644 (in Japanese with English abstract).
116. Takasu S, Tsuchiya H, Sakurai Y, et al: Acute gastric mucosal lesion (AGML) following endocopy-Its frequency and measures against it: Stomach and Intestine 1989;24:653–660 (in Japanese with English abstract).
117. Doi K: Clinical aspects of acute heterocheilidiasis of the stomach (due of the larvae of *Anisakis simplex* and *Terranova decipiens*)—Especially on it's defferential diagnosis by X-rays and endoscopy -: Stomach and Intestine 1973;8:1513–1518 (in Japanese with English abstract).
118. Karasawa Y, Takahara K, Hirafuku I, et al: Experience of anisakiasis and terranovasis of digestive tracts: Gastroenterol Eudosc 1984;26:2134–2135 (in Japanese).
119. Nagano K, Sasaki Y, Otani N, et al: Investigation of acute gastric heterocheilidiasis, especially gastic terranovasis: Int. Med. 1975;36:1030–1037 (in Japanese).
120. Koyama T, Araki J, Machida M, et al: Current problem on anisakiasis: Modern Media, 1982;28:434–443 (in Japanese).
121. Ishikura H, Kobayashi, Miyamoto K, et al: Transition of occurrence of anisakiasis and its paratenic host fishes in Japan with pathogenesis of anisakiasis: Hokkaido J Med Sci 1988;63:376–391.

122. Tsushima K, Sasamura M, Ito F, et al: A human case of terranova infection: Jpn J Parasitol. 1984;33:87 (in Japanese).
123. Chiba R, Aizawa T: Gastric anisakiasis and teranovasis in Aomori prefecture: Gastroenterol Endosc, 1988;30:2716–2717.
124. Sakaguchi Y, Annen Y, Kagei N: A first case of terranovasis found in Toyama prefecture. J. Kanagawa Univ.: 1986,3:186–189 (in Japaneses).
125. Iino H. :Occurence of anisakiais in Kyushu (6th examination) Jpn J Gastnoenterol Endosc 1985;27:630 (in Japanese).
126. Takao Y, Arita T, Emoto O: The second case of acute gastric terranovasis (larvae type A) in Kyushu: Japan: Rinsho to Kenkyu 1986;63:3636–3638 (in Japanese).

4

Antibodies to the Circumsporozoite Protein and Protective Immunity to Malaria Sporozoites

Trevor R. Jones, W. Ripley Ballou, and Stephen L. Hoffman

Malaria is caused by members of the genus *Plasmodium*. The four species infecting humans are *Plasmodium falciparum, P. vivax, P. malariae,* and *P. ovale.* Plasmodium sporozoites develop in the salivary glands of female anopheline mosquitoes and are transmitted to humans during a blood meal. The sporozoites remain in the host's circulation for a brief period before entering hepatocytes, where they develop for 5–16 days depending on the species. During this period they increase in number thousands of times. A uninucleate *P. falciparum* sporozoite, for example, can develop into a multinucleate liver schizont with as many as 30,000–50,000 uninucleate merozoites. The phase beginning with sporozoite entry into the blood and ending with merozoite invasion of erythrocytes is called the exoerythrocytic phase of the disease.

Figure 4.1. Life cycle of the malaria parasite. Sporozoites enter the blood stream when a female anopheline mosquito takes a blood meal (1). The sporozoites enter liver cells and amplify their number thousands of times through the asexual production of merozoites. When the liver cells rupture, merozoites pour into the bloodstream and enter red blood cells (2). Clinical symptoms appear when the mature erythrocytic schizont ruptures. Some of the blood stage parasites differentiate into the sexual stages, gametocytes. When these organisms are taken up by another feeding mosquito, sexual reproduction occurs within the mosquito midgut (3), and sporozoites eventually appear in its salivary glands.

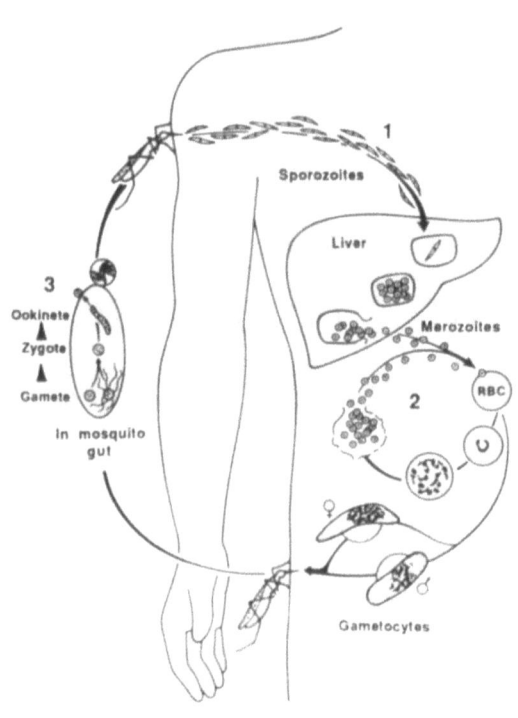

The life cycle of *Plasmodium* reveals several apparent points at which to attack and disrupt the life cycle of the malarial parasite (Fig. 4.1). Investigators are working to develop vaccines that produce immunity to circulating sporozoites, to parasites developing in the liver, to merozoites free in the bloodstream, to parasite-infected erythrocytes, and to the gametocyte. Our efforts are directed at developing vaccines to preerythrocytic stages of the parasite, in particular the sporozoite and the developing liver stage parasite.

Sporozoite Immunity and Circumsporozoite Protein

Studies with avian malarias by Richards et al. (1) indicated that immunization with inactivated sporozoites induced a partial immunity in birds. In 1967 and 1969, Nussenzweig and colleagues (2,3) demonstrated that mice immunized with radiation-attenuated *P. berghei* sporozoites were protected against challenge with infective, normal *P. berghei* sporozoites. Studies also demonstrated that this sporozoite immunity was stage-specific but not species-specific. In other words, no protection was seen when the immunized mice were challenged with blood stage parasites of *P. berghei*, but protection was seen when mice were challenged with sporozoites from either *P. vinckei* or *P. berghei*. Clyde et al. (4,5) and Rieckmann et al. (6,7) reproduced these studies in humans using irradiated *P. falciparum* and *P. vivax* sporozoites. The important demonstration that humans are biologically capable of generating a protective immune response to malarial sporozoites provided the impetus for subsequent efforts in malaria development. Studies (8,9) have shown that, at least in some strains of mice, the immunity induced by immunization with irradiated sporozoites is mediated by $CD8^+$ T-cells. A large body of evidence already exists, however, indicating that antibody-dependent immunity to sporozoites is possible, and attempts to improve methods of inducing protective antibodies have therefore continued. It now appears that there are two mechanisms of immunity to preerythrocytic stages: T-cell dependent and antibody-dependent. This chapter examines the background and current status of efforts to create vaccines that induce antibodies that prevent effective sporozoite invasion of hepatocytes.

In 1980 an apparent target of the antibody response was localized on the surface of the *P. berghei* sporozoite by Yoshida and coworkers (10). This immunogenic protein, named circumsporozoite (CS) protein, was detected through the use of a monoclonal antibody induced by the bites of irradiated *P. berghei*-infected mosquitoes. In *P. berghei*, the protein has a molecular weight of 44 kDa. Shortly thereafter, analogous proteins were found on the sporozoite surface of other species of *Plasmodium* (11,12). Within a few years of these observations, the genes for the CS proteins of *P. knowlesi* (13,14), *P. cynomolgi* (15), *P. falciparum* (16,17), *P. vivax* (18,19), *P. berghei* (20,21), *P. yoelii* (22,23), and *P. malariae* (24) were cloned and sequenced. These proteins are generally similar in structure, and all possess a highly immunogenic set of repeated amino acid sequences. The exact amino acid sequence of the repeats varies among species; some repeats are as short as four residues, others as long as twelve (Fig. 4.2).

In 1980 Potocnjak and coworkers (25) made the important observation that passive transfer of antibody alone can provide protection from sporozoite challenge. This antibody reacted with the 44-kDa CS protein that had been described by Yoshida et al. (10). Fab fragments [monovalent immunoglobulin G (IgG) fragments prepared by

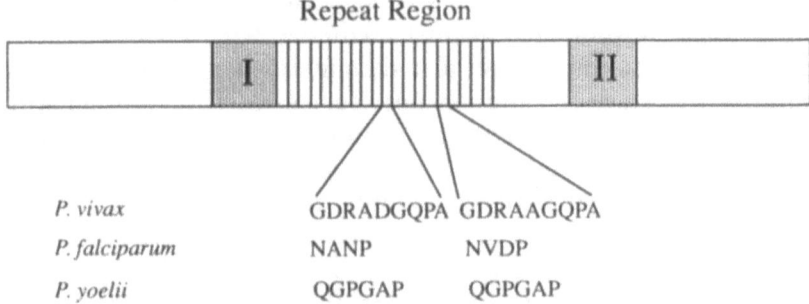

Figure 4.2. Simplified diagram of the *Plasmodium* circumsporozoite protein. Regions I and II represent highly conserved regions with considerable homology between species. The highly immunodominant central repeat region contains repeated amino acid sequences characteristic of each species. Two repeats in the diagram are expanded to show representative sequences from three species. *P. vivax* has 19 copies of a nine amino acid repeat; two of several variant sequences are shown. *P. falciparum* has 23 copies of a four amino acid repeat. NANP is the major repeat, and NVDP is the minor repeat. *P. yoelii* has 15 to 19 copies of a six amino acid repeat (QGPGAP) and six to seven copies of a minor repeat (QQPP).

papain treatment] of this antibody were then successfully used in passive transfer studies in which 10 μg of Fab were intravenously injected into mice that were protected against challenge with 1000 *P. berghei* sporozoites. In comparison studies, the Fab fragments proved as effective as intact antibody at providing protection. Within 5 years of the release of these data, the sequences of the CS proteins of several species of *Plasmodium* were published (vide supra) and a concerted effort began to develop methods of inducing antibody-mediated immunity against sporozoites.

P. falciparum and Human Vaccine Trials

With the identification of the central repeat region of the CS protein of *P. falciparum* in 1984 (16,17), attempts to induce sterilizing immunity to sporozoites began to focus on the use of peptides as immunogens. Young and colleagues (26) transformed *Escherichia coli* with the gene encoding the four amino acid repeat sequence of the *P. falciparum* CS protein; the gene product elicited high antibody titers in mice. Ballou et al. (27) used synthetic peptides based on the *P. falciparum* CS protein conjugated to a carrier to induce antibodies in mice and rabbits. These antibodies recognized native CS protein and blocked sporozoite invasion of human hepatoma cells. Antibodies raised to two conserved nonrepeat flanking regions (regions I and II) did not neutralize sporozoites. These data were interpreted to mean that the central repeat region was the preferred immunogen for a sporozoite vaccine. Weber and Hockmeyer (28) used the cloned CS protein gene from one strain of *P. falciparum* to probe 17 other *P. falciparum* strains from around the world and found that the probe hybridized with each strain. This finding led them to conclude that the gene was highly conserved and its product could be considered a good candidate for vaccine development. Zavala et al. (29) used sera from a malarious area, monoclonal antibodies to sporozoites, and polyclonal antibodies to (NANP)$_3$ to show that antibody raised against the native immunogens reacted with (NANP)$_3$, and antibody to (NANP)$_3$ reacted with sporozoites. This finding also supported the contention that the central repeat region was an excellent peptide on which to base a sporozoite vaccine. Zavala et al. (30) and Yoshida and colleagues (31) further showed that *P. falciparum* sporozoites collected worldwide reacted with antibody to the repeat region of the CS protein, thereby implying that there was no antigenic variation among strains, and that the repeat region was an excellent target for development as a vaccine.

The first human vaccines against the *P. falciparum* sporozoite were based on the sequence NANP and the minor repeat NVDP. Ballou and colleagues (32) used a recombinant protein vaccine containing copies of both NANP and NVDP plus 32 amino acids from a bacterial tetracycline resistance gene and adsorbed this vaccine to alum (R32tet$_{32}$). Herrington and coworkers (33) used a synthetic peptide vaccine consisting of three copies of NANP conjugated to tetanus toxoid and also adsorbed to alum. In the Ballou study, six subjects immunized with 100-to 800-µg doses were challenged with the bites of five infected mosquitoes; one subject was protected. In the Herrington study, three vaccinated subjects were likewise challenged, and one was protected. In both studies the subjects not infected upon challenge were also the ones with the highest antibody levels, and five of nine of these subjects experienced lengthened prepatency periods that correlated with the antibody titer. The antibody levels induced by these vaccines were low in most volunteers, although the few subjects who responded well developed titers that approached those in persons from malaria-endemic areas. As Hoffman and coworkers (34) demonstrated, however, naturally acquired antibodies to sporozoites seen in persons living in malarious areas do not correlate with resistance to malaria.

Antibody Induction with Subunit Vaccines

The initial human vaccine trials highlighted two issues concerning CS repeat vaccines. The first was the relatively poor antibody response of a significant number of individuals to the vaccines, and the second was the poor efficacy in the presence of what would be considered relatvely high antibody titers for vaccines against viruses or bacteria. During this period, the CS protein of the murine malaria parasite P. berghei was cloned and sequenced (20,21). To evaluate the response to subunit CS vaccines, Egan and colleagues (35) immunized mice with a peptide consisting of an eight amino acid repeat sequence found in the CS protein of P. berghei (DPAPPNAN conjugated to keyhole limpet hemocyanin). Other mice received an E. coli produced recombinant protein containing a large portion of the CS protein including the entire repeat region. By using complete and incomplete Freund's adjuvant, they were able to achieve high levels of antibody, but both vaccines protected only about 50% of mice challenged with a small number of infective sporozoites. Naive mice were then passively immunized with varying amounts of the purified IgG from the immunized mice, and protective efficacy was compared to that achieved with a protective monoclonal antibody. All mice receiving the monoclonal antibody were protected; and 75% were protected with purified polyclonal IgG taken from mice immunized with peptide. These data implied that, on a weight-to-weight basis, polyclonal antibodies appeared at least as effective at providing protection as did monoclonal antibodies. Later, in 1987, Zavala and colleagues (36) obtained even better results by immunizing with a discrete protein. They immunized with a slightly different repeat (DPPPPNPN conjugated to tetanus toxoid) from the P. berghei CS protein than the one uses by Egan et al. (35). The vaccinated mice were challenged with a small number (1000) of infective sporozoites. This number is considered a small challenge because the dose of P.berghei sporozoites required to infect 50% of mice (ID_{50}) is generally about 500 sporozoites and because P. berghei irradiated sporozoites induce an immune state so solid that mice can withstand sporozoite challenges of 50,000–500,000 infective sporozoites (37; S.L. Hoffman and W.R. Ballou; unpublished data). Nevertheless, up to 87% of the mice were protected.

Charoenvit and colleagues (38) showed that antibody reactivity with sporozoite proteins did not imply the ability to protect. Five monoclonal antibodies generated against P. yoelii sporozoites were positive for reactivity to P. yoelii sporozoites by immunofluorescence; only three were positive in the circumsporozoite precipitation test. Of these three, only two provided protection when sporozoites were incubated with antibody prior to injection into naive mice (sporozoite neutralization technique). Shortly after this work was published, the sequence of the *P. yoelii* CS protein was determined (22,23) using one of the protective antibodies (NYS1) developed by Charoenvit et al. (38). The protein contained a central repeat region with a dominant repeat having the sequence QGPGAP and the minor repeat QQPP. Incubation with synthetic QGPGAP inhibited the reactivity of the protective antibody NYS1. Sedegah and coworkers (39) used either irradiated *P. yoelii* sporozoites or a vaccinia recombinant construct encoding the full-length *P. yoelii* CS protein including 19 copies of QGPGAP to vaccinate mice. Both groups of animals generated excellent antibodies against the (QGPGAP)$_2$; the irradiated sporozoite vaccinated group was protected upon challenge with 10,000 infective sporozoites, the recombinant construct vaccinated group was not protected against challenge with 200 sporozoites. However, when the protected mice were depleted of CD8$^+$ T-cells, protection was lost, indicating that neither group generated protective antibodies. Charoenvit and colleagues (40) showed that mice that received NYS1, a monoclonal antibody to the *P. yoelii* CS protein repeat region, were protected upon subsequent sporozoite challenge. However, mice immunized with a *P. yoelii* repeat region subunit vaccine were not protected. Upon examination of the sera from the actively immunized, passively immunized (NYS1 antibody), and irradiated sporozoite immunized mice, no significant difference between the three types of sera was found by the enzyme-linked immunosorbent assay (ELISA) against QGPGAP or immunofluorescence against sporozoites.

These studies, viewed collectively, demonstrate that (1) circulating antibodies alone (in the form of injected monoclonal antibodies) provide potent protection against sporozoite challenge; and (2) polyclonal antisera generated by vaccination with a protein containing the target of the protective antibody provided only partial or no antibody-mediated protection, and the mice remained susceptible to sporozoite challenge. These experiments left two important questions unanswered. Can peptide vaccines induce a protective monoclonal antibody? (All protective monoclonal antibodies to date had been raised by immunization with irradiated sporozoites.) Can peptide vaccines induce a solid, protective polyclonal response as effective as that induced by passive immunization with monoclonal antibodies or active immunization with irradiated sporozoites?

Antibody-Mediated Immunity and Specific Epitopes

Ak and coworkers (manuscript in preparation) used a *P. yoelii* repeat region subunit vaccine to immunize mice and make monoclonal antibodies. Two of the antibodies, an IgG1 and an IgG2b, provide protection upon passive transfer. These findings show that the immune system, at least in mice, is capable of recognizing a small peptide sequence and generating a monoclonal antibody against it that is protective. The issue of whether antibody-mediated protection is IgG subclass-dependent was also put to rest. These data, in combination with the Fab studies of Potocnjak and colleagues (25), show that IgG subclass is not a restricting factor in the ability to provide protection. In the *P. yoelii* system, IgG1, IgG2b, and IgG3 monoclonal antibodies have provided protection (NYS1, a protective monoclonal antibody is an IgG3). In 1984 Charoenvit and colleagues (unpublished data) generated a monoclonal antibody by immunizing mice with irradiated *P. vivax* sporozoites. This antibody, designated NVS3, was used in the cloning and sequencing of the *P. vivax* CS protein, and it binds to the repeat region (DRA A/D GQPAG) of that protein (19). Work by Charoenvit and coworkers (41) demonstrated that when 2 mg of NVS3 were infused into *Saimiri* monkeys, four of six of these monkeys were protected against a 10,000 *P. vivax* sporozoite challenge. This study was the first demonstration that sporozoites from a human malarial parasite could be neutralized solely by circulating antibodies. NVS3 was then subjected to analysis by epitope mapping (42), which required synthesis of many eight amino acid subsets of the repeat region of the *P. vivax* CS protein and showed clearly that the epitope of NVS3 was the four amino acid sequence AGDR (41). In an earlier study performed by Collins et al. (43), *Saimiri* monkeys were immunized with a recombinant subunit vaccine designated $NS1_{81}V20$. This vaccine contains multiple copies of the repeat sequence of the *P. vivax* CS protein and of AGDR. The monkeys immunized with $NS1_{81}V20$ were not protected upon challenge with 10,000 *P. vivax* sporozoites. Subsequent examination of the sera from these monkeys (41) indicated that although they generated high antibody titers to *P. vivax* sporozoites by immunofluorescence and to $NS1_{81}V20$ in ELISA none of the animals produced detectable antibody to $(AGDR)_2$, as determined in ELISA even though multiple copies of AGDR are contained within the $NS1_{81}V20$ sequence. This finding means that even if the desired protective epitope is present in the sequence of the immunogen, the desired antibody response is not ensured. Extraneous amino acids that form nonprotective but apparently immunodominant epitopes may have to be deleted from the immunogen. These studies demonstrated that great exactitude is required in the selection of the epitope used to induce a protective antibody response. Here an immunogen that is bound by a monoclonal antibody known to be protective in passive transfer studies generated an excellent antibody titer against sporozoites and against itself but failed to produce antibody to the discrete, protective epitope contained within its sequence.

The promise of the AGDR epitope is tempered, however, when one considers data on strain variation in the CS protein repeat region of *P. vivax*. Zavala and coworkers (30) studied seven strains of *P. vivax* from around the world and found that monoclonal antibodies to the repeat region reacted with sporozoites from all seven strains. This nonvariability was consistent with that found in *P. falciparum*. In 1989, however, Rosenberg and colleagues (44) found that more than 14% of the uncomplicated cases of *P. vivax* malaria at two sites in Thailand were caused by a strain with a nine amino acid repeat that shared only a three amino acid homology with the previously published, nonvariant sequences. Antibody reactivity to the variant repeat in persons from *P. vivax* endemic areas has since been reported (45,46).

This variant sequence does not contain AGDR. The possibility, however, that AGDR can be used to immunize and provide protection against nonvariant strains of *P. vivax* means that it may be a paradigm for the development of synthetic or recombinant vaccines against a variety of CS protein repeats in both *P. vivax* and other species of *Plasmodium*.

Carriers and Adjuvants

Concurrent with advances in understanding the crucial importance of epitope specificity (vide supra), methods for significantly improving the antibody titers induced by subunit vaccines are also being developed. Changes in both carriers and the adjuvants have resulted in dramatic improvements in antibody response. J.C. Sadoff and colleagues (Walter Reed Army Institute of Research, personal communication) chemically conjugated R32 (32 copies of the *P. falciparum* repeats) to a variety of protein carrier molecules including tetanus toxoid, choleragenoid, meningiococcal outer membrane protein, diphtheria, and the exotoxin A of *Pseudomonas aeruginosa* (ToxA). All were safe when used in humans, but the R32ToxA construct was the superior immunogen, eliciting levels of antibodies in nearly all volunteers equal to those found in the protected volunteers in the Ballou and Herrington studies (32,33). In addition to improved carriers, new and more effective adjuvants are also now available. Using R32NS1$_{81}$ (R32 plus 81 amino acids from nonstructural protein 1 of influenza A) as the vaccine/carrier, Rickman et al. (47) have shown that the use of a novel adjuvant consisting of detoxified lipid A (MPL) and mycobacterial cell wall skeleton (CWS) in squalane (Detox, Ribi Immunochem) results in as much as a 10-fold increase in immunogenicity compared to R32NS1$_{81}$ adsorbed to alum, the traditional adjuvant. Similar results have been observed when the R32NS1$_{81}$ antigen was incorporated into liposomes containing detoxified lipid A (C. Alving and L. Fries, personal communication). Efficacy studies using these new carrier/adjuvant combinations are currently under way.

Longevity of Antibody

For vaccines designed to induce antibody-mediated protection, the longevity of antibody levels to specific, protective epitopes on the malaria parasite is an important factor when determining the duration of protection. Persons such as travelers who pass through malarious areas require protection for only a specified period of time, whereas persons living permanently in endemic areas need protective antibody levels for life. Antibody titers induced by R32 vaccine candidates combined with both alum and MPL/CWS adjuvants generally drop to 50% with 6 months of the final vaccination (32,47). The effect of this drop on efficacy is unknown. There are several possible ways to extend antibody longevity. Adjuvants or other drug delivery methods may be developed that lead to the continuous production of specific antibody over long periods of time. Another approach would be to include a T-helper epitope in the vaccine. Persons living in endemic areas may receive natural T-helper cell-mediated boosting owing to exposure to the bites of infected mosquitoes. In rodents, there is evidence that sporozoites can boost the level of antibody to subunit vaccines (48), but whether natural exposure can boost or even help maintain antibody levels in humans is not known.

Antibody-Mediated Passive Immunization

A variety of persons including tourists, diplomats, businessmen, and military personnel may receive effective levels of protection from passively transferred antibody. Studies with rodent and human malarias, already described (25,41), demonstrated that circulating antibodies alone can induce protection. The technology exists to convert murine monoclonal antibodies to chimeric human immunoglobulin (49), thereby reducing reactivity due to species differences. This reaction may provide an avenue for the production of injectable antibody capable of providing short-term protection to those briefly exposed to malaria. In another approach, lymphocytes from appropriately immunized humans are fused with myeloma cells to make a hybridoma that produces fully human monoclonal antibody. It is also possible to produce vast numbers of different Fab fragments by expression in E. coli and screen them for reactivity to any desired epitope (50). Once the desired Fab fragment is identified, it can be produced in great quantity. Because Fab fragments have been shown to be protective, at least in mice (25), it is conceivable that they could then be used for passive immunization.

Overview

Today's efforts to develop a sporozoite vaccine were started with the observation that irradiated sporozoites induced protection in rodents. With the identification and sequencing of the circumsporozoite protein, attention focused on the creation of specific synthetic or recombinant subunit molecules that would induce a potent immunity to sporozoites. Passive transfer experiments using monoclonal antibodies generated against both irradiated sporozoites and peptides demonstrate that impressive levels of protection against sporozoite challenge can be achieved with circulating antibodies alone. The next step, that of obtaining similar antibody-mediated protection through immunization with subunit vaccines, has not been reached. Information obtained from both P. vivax and P. yoelii studies, however, indicates that the successful induction of protection with monoclonal antibodies depends on the exact specificity of the antibody. The problems encountered in inducing polyclonal antibody-mediated protection that is equivalent to passive immunization with protective monoclonal antibodies suggests that polyclonal responses are perhaps too diffused and of too low titer to neutralize the sporozoite.

Two important factors in the induction of antibody-mediated sporozoite immunity have emerged. First, methods for inducing and maintaining high levels of antibodies must be developed. MPL/CWS adjuvants and liposomes have already been shown to induce those high levels. Methods for increasing the longevity of antibody, however, have yet to be developed. Second, it may be ideal for the immunogen to contain only epitope (s) proved to be protective when bound by specific antibody. These epitopes can probably be best identified by the generation of monoclonal antibodies to promising epitopes and subsequent passive transfer studies. Preparing an epitope-carrier-adjuvant combination that induces high titer antibodies with extended longevity and correct specificity is now the focus of interest for many malaria vaccine developers.

Acknowledgments

This study was supported by Naval Medical Research and Development Command work unit numbers 3M16102BS13AK111, 3M162787AB70AN121, and 3M463807D808AQ133. Those experiments reported herein that were performed by Department of Defense personnel were conducted according to the principles set forth in the *Guide for the Care and Use of Laboratory Animals*, Institute of Laboratory Animal Resources, National Research Council, DHHS Publication (NIH) 86-23, (1985). The opinions and assertions contained herein are those of the authors and are not to be construed as official or as reflecting the views of the Navy Department or the Army Department.

References

1. Richards WHR: Active immunization of chicks against Plasmodium gallinaceum by inactivated homologous sporozoites and erythrocytic parasites. Nature 1966;212:1492–1494.
2. Nussenzweig RS, Vanderberg J, Most H, Orton C:. Protective immunity produced by the injection of x-irradiated sporozoites of Plasmodium berghei. Nature 1967;216:160–162.
3. Nussenzweig RS, Vanderberg J, Most H, Orton C:. Specificity of protective immunity produced by x-irradiated Plasmodium berghei sporozoites. Nature 1969;222:488–489.
4. Clyde DF, Most H, McCarthy VC, Vanderberg JP:. Immunization of man against sporozoite-induced falciparum malaria. Am J Med Sci 1973; 266:169–177.
5. Clyde DF, McCarthy VC, Miller RM, Woodward WE:. Immunization of man against falciparum and vivax malaria by use of attenuated sporozoites. Am J Trop Med Hyg 1975;24:397–401.
6. Rieckmann KH, Carson PE, Beaudoin RL, et al: Sporozoite induced immunity in man against an ethiopian strain of Plasmodium falciparum. Trans R Soc Trop Med Hyg 1974; 68:258–259.
7. Rieckmann KH, Beaudoin RL, Cassells JS, Sell KW: Use of attenuated sporozoites in the immunization of human volunteers against falciparum malaria. Bull WHO 1979; 57(suppl): 261–265.
8. Schofield L, Villaquiran J, Ferreira A, et al:. γ Interferon, CD8$^+$ T cells and antibodies required for immunity to malaria sporozoites. Nature 1987;330:664–666.
9. Weiss WR, Sedegah M, Beaudoin, RL, et al:. CD8$^+$ T cells (cytotoxic/

suppressors) are required for protection in mice immunized with malaria sporozoites. Proc Natl Acad Sci USA 1988;85:573–576.
10. Yoshida N, Nussenzweig RS, Potocnjak P, et al: Hybridoma produces protective antibodies directed against the sporozoite stage of malaria parasite. Science 1980;207:71–73.
11. Nardin EH, Nussenzweig V, Nussenzweig RS, et al: Circumsporozoite proteins of human malaria parasites Plasmodium falciparum and Plasmodium vivax. J Exp Med 1982;156:20–30.
12. Santoro F, Cochrane AH, Nussenzweig V, et al: Structural similarities among the protective antigens of sporozoites from different species of malaria parasites. J Biol Chem 1983;258:3341–3345.
13. Ellis J, Ozaki LS, Gwadz RW, et al: Cloning and expression in E. coli of the malarial sporozoite surface antigen gene from Plasmodium knowlesi. Nature 1983; 302:536–538.
14. Godson GN, Ellis J, Svec, P, et al: Identification and chemical synthesis of a tandemly repeated immunogenic region of Plasmodium knowlesi circumsporozoite protein. Nature 1983;305:29–33.
15. Enea V, Arnot D, Schmidt EC, et al: Circumsporozoite gene Plasmodium cynomolgi (Gombak): cDNA cloning and expression of the repetitive circumsporozoite epitope. Proc Natl Acad Sci USA 1984; 81:7520–7524.
16. Dame JB, Williams JL, McCutchan TF, et al: Structure of the gene encoding the immunodominant surface antigen on the sporozoite of the human malaria parasite Plasmodium falciparum. Science 1984;225:593–599.
17. Enea V, Ellis J, Zavala F, et al: DNA cloning of Plasmodium falciparum circumsporozoite gene: amino acid sequence of repetitive epitope. Science 1984;225:628–630.
18. Arnot DE, Barnwell JW, Tam JP, et al: Circumsporozoite protein of Plasmodium vivax: gene cloning and characterization of the immunodominant epitope. Science 1985;230:815–818.
19. McCutchan TF, Lal AA, de la Cruz VF, et al: Sequence of the immunodominant epitope for the surface protein on sporozoites of Plasmodium vivax. Science 1985;230:1381–1383.
20. Eichinger DJ, Arnot DE, Tam, JP, et al: Circumsporozoite protein of Plasmodium berghei: gene cloning and identification of the immunodominant epitopes. Mol Cell Biol 1986;6:3965–3971.
21. Weber JL, Egan JE, Lyon JA, et al: Plasmodium berghei: cloning of the circumsporozite protein gene. Exp Parasit. 1987; 63:295–300.
22. Lal AA, de la Cruz VF, Welsh JA, et al: Structure of the gene encoding the circumsporozoite protein of Plasmodium yoelii. J Biol Chem 1987; 262:2937–2940.
23. Teisin T, Gross M, Young JF, et al: Cloning and sequencing of the Plasmodium yoelii circumsporozoite gene. Fed Proc 1987;46:2185 (abstract).
24. Lal AA, de la Cruz VF, Campbell GH, et al: Structure of the circumsporozoite gene of Plasmodium malariae. Mol Biochem Parasit 1988;30: 291–294.
25. Potocnjak P, Yoshida N, Nussenzweig RS, Nussenzweig V: Monovalent fragments (Fab) of monoclonal antibodies to a sporozoite surface antigen

(Pb44) protect mice against malarial infection. J Exp Med 1980;151:1504–1513.
26. Young JF, Hockmeyer WT, Gross M, et al: Expression of Plasmodium falciparum circumsporozoite proteins in Escherichia coli for potential use in a human malaria vaccine. Science 1985;228:958–962.
27. Ballou WP, Rothbard J, Wirtz RA, et al: Immunogenicity of synthetic peptides from circumsporozoite protein of Plasmodium falciparum. Science 1985;228:996–999.
28. Weber JL, Hockmeyer WT: Structure of the circumsporozoite protein gene in 18 strains of Plasmodium falciparum. Mol Biochem Parasitol 1985;15:305–316.
29. Zavala F, Tam JP, Hollingdale MR, et al: Rationale for development of a synthetic vaccine against Plasmodium falciparum malaria. Science 1985;228:1436–1440.
30. Zavala F, Masuda A, Graves PM, et al: Ubiquity of the repetitive epitope of the CS protein in different isolates of human malaria parasites. J Immunol 1985;135:2790–2793.
31. Yoshida N, del Portillo HA, di Santi SM, et al: Plasmodium falciparum: epidemiological studies on the circumsporozoite gene. Exp Parasitol 1987;64:510–513.
32. Ballou WP, Hoffman SL, Sherwood JA, et al: Safety and efficacy of a recombinant DNA Plasmodium falciparum sporozoite vaccine. Lancet 1987;1:1277–1281.
33. Herrington DA, Clyde DF, Losonsky G, et al: Safety and immunogenicity in man of a synthetic peptide malaria vaccine against Plasmodium falciparum sporozoites. Nature 1987;328:257–259.
34. Hoffman S, Oster CN, Plowe CV, et al: Naturally acquired antibodies to sporozoites do not prevent malaria: vaccine development implications. Science 1987;237:639–642.
35. Egan JE, Weber JL, Ballou WR, et al: Efficacy of murine malaria sporozoite vaccines: Implications for human vaccine development. Science 1987;236:453–456.
36. Zavala F, Tam JP, Barr PJ, et al: Synthetic peptide vaccine confers protection against murine malaria. J Exp Med 1987;166:1591–1596.
37. Beaudoin RL, Strome CPA, Mitchell F, Tubergen TA: Plasmodium berghei: immunization of mice against the ANKA strain using the unaltered sporozoite as an antigen. Exp Parasitol 1977;42:1–5.
38. Charoenvit Y, Leef MF, Yuan LF, et al: Characterization of Plasmodium yoelii monoclonal antibodies directed against stage-specific sporozoite antigens. Infect, Immun 1987;55:604–608.
39. Sedegah M, Beaudoin RL, de la Vega P, et al: Use of a vaccinia construct expressing the circumsporozoite protein in the analysis of protective immunity to Plasmodium yoelii. In Lasky L (ed): 1988 Technological Alzn R. Liss, Inc, New York Advances in Vaccine Development. pp. 295–309.
40. Charoenvit Y, Mellouk S, Cole C, et al: Monoclonal, but not polyclonal antibodies protect against Plasmodium yoelii sporozoites. J Immunol 1991;146:1020–1025.

41. Charoenvit Y, Collins WE, Jones TR, et al: Inability of malaria vaccine to induce antibodies to a protective epitope contained within its sequence. Science, 1991;251:668–671.
42. Geysen HM, Rodda SJ, Mason TJ, et al: Strategies for epitope analysis using peptide synthesis. J Immunol Methods 1987;102:259–274.
43. Collins WE, Nussenzweig RS, Ballou WR, et al: Immunization of Saimiri scireus boliviensis with recombinant vaccines based on circumsporozoite protein of Plasmodium vivax. Am J Trop Med Hyg 1989;40: 455–464.
44. Rosenberg R, Wirtz RA, Lanar DE, et al: Circumsporozoite protein heterogeneity in the human malaria parasite Plasmodium vivax. Science 1989;245:973–976.
45. Wirtz RA, Rosenberg R, Sattabongkot J, Webster HK: Prevalence of antibody to heterologous circumsporozoite protein of Plasmodium vivax in Thailand. Lancet 1990;336:593–595.
46. Cochrane AH, Nardin EH, de Arruda M, et al: Widespread reactivity of human sera with a variant repeat of the circumsporozoite protein of Plasmodium vivax. Am J Trop Med Hyg 1990;43:446–451.
47. Rickman LS, Gordon DM, Wistar R, et al: Use of adjuvant containing mycobacterial cell-wall skeleton, monophosphoryl lipid A, and squalane in malaria circumsporozoite protein vaccine. Lancet 1991;337:998-1001.
48. Hoffman SL, Cannon LT, Berzofsky JA, et al: Plasmodium falciparum: sporozoite boosting of immunity due to a T-cell epitope on a sporozoite vaccine. Exp Parasitol 1987;64:64–70.
49. Morrison SL, Johnson MJ, Herzenberg LA, Oi VT: Chimeric human antibody molecules: Mouse antigen-binding domains with human constant region domains. Proc Natl Acad Sci USA 1984;81:685–6855.
50. Huse WD, Sastry L, Iverson SA, et al: Generation of a large combinatorial library of the immunoglobulin repertoire in phage lambda. Science 1989;246:1275–1281.

5
Human T-Cell Responses in *Leishmania* Infections

Donna M. Russo, Manoel Barral-Netto, Aldina Barral, and Steven G. Reed

Leishmaniasis is a major public health problem in several areas of the world, occurring in large areas of the Middle East, Africa, and Latin America. The *Leishmania* complex is a group of closely related parasites that occupy a wide variety of ecologic niches and cause a spectrum of clinical disease. Infection is initiated by the bite of a sandfly, which injects the motile form, called promastigotes. Once inside the mammalian host, *Leishmania* are obligatory intramacrophage parasites, multiplying in these cells as nonmotile amastigotes. Basic biologic aspects of the parasite are thoroughly reviewed elsewhere (1,2).

Human infections are generally manifested as acute or subclinical visceral infections, localized cutaneous (single or multiple) lesions, or mucosal disease of varying severity. The clinical manifestations of tegumentary leishmaniasis include single cutaneous ulcerations (CL = cutaneous leishmaniasis, most commonly caused by *Leishmania major* or *L. amazonensis*), which heal spontaneously or after chemotherapy; severely disfiguring mucosal involvement (ML = mucosal leishmaniasis, most commonly caused by *L. braziliensis*), which is often refractory to chemotherapy; and anergic diffuse cutaneous leishmaniasis (DL) characterized by the absence of a cellular immune response. Instead of tegumentary involvement, *Leishmania* preferentially multiply in the bone marrow, spleen, and liver, leading to a range of subclinical to clinical infections, the most severe form of which is acute visceral leishmaniasis (VL, most commonly caused by *L. donovani* or *L. chagasi*). This acute form of the disease is characteristically fatal if untreated. Despite the variability of types of associated pathology, there appears to be at least one feature common among the human leishmaniases: recovery from and resistance to disease are dependent on protective T-cell responses. Furthermore, upon clinical recovery from most forms of leishmaniasis there is solid and long-lasting immunity to reinfection.

There is a solid basis, in both theory and practice, that lasting immunity against leishmaniasis can be specifically induced. The history of *Leishmania* immunology has roots in traditional folk medicine in which material from active lesions was inoculated into naive recipients, often in the buttocks (3). This practice produced a lesion at a controlled site, and once this lesion self-cured the inoculated individual would be immune to a more disfiguring facial lesion. This simple but effective practice illustrates two important hallmarks of T-cell-mediated immunity: It is often most effectively induced with a live preparation, and it is long-lasting. Modern immunization efforts are somewhat more sophisticated (reviewed in ref. 1) but still consist of whole-parasite preparations. For most types of leishmaniasis, immunization of humans has not been possible. The only real success has been for Old World cutaneous leishmaniasis using live vaccines—in part due to the nature of these infections, which are generally localized and self-limiting with long-lasting associated immunity after recovery. Clinically similar leishmaniasis occurs in Latin America, and there is good indication that vaccination against this type of disease can be successful there as well. Furthermore, animal studies and observations in patients recovered from acute disease have indicated that immunization against most or all forms of leishmaniasis is possible.

Cellular Responses of Leishmaniasis Patients

Delayed Hypersensitivity Responses

Skin testing for leishmaniasis has been a valuable clinical and epidemiologic tool for decades. Montenegro (4) reported the use of promastigote extract to elicit delayed hypersensitivity (DH) responses in patients with cutaneous leishmaniasis in 1926. Since then, a variety of crude antigen preparations have been used, all promastigote derived, for the *Leishmania* skin test. Most commonly used are whole, fixed promastigotes. Although the term "Montenegro" test is often applied to the use of such preparations, he originally used soluble extracts. More recent work has shown that soluble promastigote extracts are generally superior to intact promastigotes (5,6), but defined antigens have not yet been used to elicit DH responses in leishmaniasis.

Patients with active cutaneous or mucosal leishmaniasis have positive DH responses to *Leishmania*, and these responses remain active indefinitely after clinical cure (5). The intensity of the responses varies, influenced largely by the duration of infection, but the highest responsiveness is found with active mucosal disease. This observation, coupled with the unusually strong in vitro lymphocyte responses in these patients (7,8), suggests a state of hypersensitivity in ML. Both Old World and New World VL are characterized by negative responses during active infections that convert to positive following clinical cure (5,9–11). With both CL and VL, positive DH responses are directly correlated with resistance to future disease.

This correlation is useful for prognosis. With VL, conversion from a negative to a positive response indicates successful therapy. This point may be particularly useful for evaluating new treatment procedures. Another application of the skin test is in epidemiologic studies. From a 12-year prospective study in a leishmaniasis endemic area we have characterized subclinical VL as the most common manifestation of infection (12). Thus only a small percentage of individuals who contract *L. chagasi* actually develop acute disease. Most infected individuals, detected by serology, resolve their infection as indicated by the development of a positive skin test. These individuals are considered to be free of risk from acute disease. Immunologic evaluation of a group of children with subclinical infection suggests an association between the development of disease and the absence of DH to *Leishmania* antigen during the early phase of infection. Children with strong proliferative responses to *Leishmania* antigen and who exhibit a positive *Leishmania* skin test do not progress to acute VL. Finally, the development of a positive DH response to *Leishmania* antigen has been used to evaluate the success of immunization trials. Conversion to a positive response is sometimes used as a parameter of successful immunization (13).

In Vitro Lymphocyte Responses

The most compelling evidence for the importance of T-cells in leishmaniasis continues to be in vitro and in vivo *Leishmania*-specific responses of patient lymphocytes. In general the in vitro lymphocyte responses of individuals with subclinical, active, or cured leishmaniasis corroborate observations of skin test responsiveness. Thus active or healed cutaneous leishmaniasis is characterized by strong DH responses and parasite-driven in vitro T-cell stimulation. These responses may persist long after specific antibody titers have fallen to or near background levels. With VL, patients lack *Leishmania*-specific DH responses during acute disease, a time when specific antibody titers are high, and their lymphocytes fail to proliferate in vitro. After resolution of symptoms, in vitro lymphocyte responses are positive (14). Although leukopenia occurs in patients with acute VL, there may or may not be a significant reduction in circulating lymphocyte numbers, indicating that the depressed responsiveness is not due merely to the lack of available lymphocytes. In contrast to the absence of proliferative responses (14) and cytokine production (15) by patient's peripheral blood leukocytes exposed to *Leishmania* antigens, these cells may produce cytokines when stimulated with mitogens (15). It is possible that antigen-specific lymphocytes become sequestered in the liver and spleen during acute infection, thus being unavailable to the circulation. T-cells obtained from treated patients regain their capacity to produce cytokines in response to *Leishmania* antigens (15–17).

The T-cell unresponsiveness during active disease may be due to a defect in accessory cell function, defective T-cell function, or both. Another possible explanation for the inability of cells to respond to *Leishmania* antigen could be the existence of suppressive cells or soluble products (18,19) inhibiting otherwise responsive cells. Peripheral blood lymphocytes from acute VL patients inhibited the mitogen-driven interleukin-2 (IL-2) production of autologous cells obtained after recovery of the patient (17). Similarly, there is evidence for the presence in acute VL patients of circulating T-cells capable of suppressing antigen-driven lymphocyte proliferation (20). Understanding the mechanisms of immune suppression in VL may help improve management of the disease. VL patients have both an inability to control the replication of *Leishmania* in macrophages and to maintain normal immune function. Thus there are fundamental reasons to emphasize approaches that restore immune competence in these individuals. To date, attempts to restore *Leishmania*-specific responses in vitro have been unsuccessful. For example, addition of IL-2 to cultures of peripheral blood mononuclear cells (PBMC) has not restored proliferative responses (16).

A different immunologic picture is associated with CL. Most CL patients respond well to antimony therapy, and self-healing is occasionally observed. Strong DH responses and in vitro proliferative responses are usually associated with cure, although patients with large ulcerated lesions generally have strong in vivo and in vitro cellular responses as well. ML is much more difficult to define and to manage. It is frequently refractory to treatment and manifests as a persistent or recurrent disease. The pathogenesis of the mucosal lesions that occur is not understood. With active ML, the intradermal skin test and lymphocyte proliferative responses are strong (7,8), and lymphocytes from ML patients can produce interferon-γ (IFN-γ), which inhibits intracellular replication of Leishmania (8). It is possible that ML represents a hypersensitivity reaction to Leishmania, which may explain in part some of the features of the disease such as the destructive attack on host tissue and the paucity of parasites in ML lesions.

The importance of cellular responses is also demonstrated in the histopathologic picture of tegumentary leishmaniasis. Fibrinoid necrosis, granuloma formation, and fibrosis have been correlated with protection status in humans (21). More detailed investigations of lymphocyte populations in tegumentary lesions have demonstrated both $CD4^+$ and $CD8^+$ cells (22), T-cells expressing IFN-γ, and T-cells bearing the "memory phenotype" (23).

The observations regarding in vivo and in vitro lymphocyte responsiveness in the various types and stages of Leishmania infection document the importance of T-cell responses in controlling disease. From the reports of successful immunization against clinical and experimental leishmaniasis and the strong correlation between immune status and T-cell responsiveness, it is apparent that efforts should be directed at identifying Leishmania antigens that induce and elicit T-cell responses. This area is one of intense investigation and is discussed below.

Lymphocyte Responses in Leishmaniasis

T-cells appear to have two main mechanisms for coping with intracellular pathogens: (1) host cell activation, e.g., activation of infected macrophages to kill their intracellular parasites; and (2) the ability to lyse infected target cells. These two effector mechanisms are linked to two pathways of antigen presentation and appear to be mediated by phenotypically distinct T-cell subsets. T-cells expressing the CD4 molecule recognize exogenous antigens processed by the endosomal pathway in association with MHC class II molecules. These cells have traditionally been ascribed a T-helper function; i.e, they stimulate other cells by cytokine secretion. Many pathogens, including Leishmania, reside and proliferate inside nonactivated mononuclear phagocytes. Such cells can be activated by cytokines produced by T-cells, including IFN-γ (24,25) and granulocyte macrophage colony stimulating factor (GM-CSF) (26), thus acquiring antimicrobial intracellular capacity. It may be due to the increased capacity of the macrophage to produce toxic oxygen (24) and nitrogen (27,28) radicals or to other as yet undefined mechanisms.

The CD4 population of murine T-cells has been further divided into two functional subsets, Th1 (T helper 1) and Th2 cells (29,30), based on their profile of cytokine secretion. Th1 cells secrete IFN-γ and IL-2, and they mediate cellular immune responses including DH reactions. Th2 cells produce IL-4 and IL-5 and enhance primarily humoral responses. The selective activation of these T-cell populations was shown to regulate cutaneous leishmaniasis in mice caused by infection with *Leishmania major*. Scott et al. (31) demonstrated that an antigen-specific T-cell line that protected BALB/c mice from fatal infections produced IFN-γ and IL-2 in response to stimulation with crude *L. major* antigen or mitogen. This profile of cytokine production corresponds to the Th1 subset of CD4$^+$ T-lymphocytes. In contrast, a T-cell line generated with a different fraction of soluble antigen did not confer protection but led to exacerbation of disease. The nonprotective line responded to antigen or mitogen stimulation by producing IL-4 and IL-5, placing it in the Th2 subset. The correlation between differential patterns of cytokine secretion and protection was further supported by the observation that IFN-γ mRNA was found in spleens and lymph nodes of *L. major*-resistant C57BL/6 mice during infection, whereas IL-4 mRNA appeared in these tissues of susceptible BALB/c mice (32). In addition, the administration of anti-IL-4 to BALB/c mice substantially delayed the progression of disease (32). Therefore the effectiveness of T-cell responses to *Leishmania* antigens may depend at least in part on the types of T-cells that are stimulated to responsiveness or nonresponsiveness. As yet, the ability of human T-cells to respond to defined *Leishmania* antigens by distinct patterns of cytokine production has not been demonstrated.

CD8$^+$ T-cells primarily recognize endogenously processed antigens that bind class I molecules intracellularly and are transported to the cell surface. CD8$^+$ T-cells frequently have cytotoxic function, but they may also secrete cytokines, notably IFN-γ. They may play a role in immunity to intracellular pathogens by either of these mechanisms. The role of cytotoxic T-cells in *Leishmania* infection has not been extensively examined, but evidence suggests that cytotoxicity is an important effector mechanism in the immune response to other nonviral intracellular pathogens, including *Listeria* (33) and *Plasmodium* (34).

T-Cell Subsets in Human Leishmaniasis

In fact, little is known regarding the types of T-cells that respond to parasite antigens during infection, including populations that may be responsible for disease resolution or those that contribute to disease progression. Melby et al. (35,36) demonstrated that CD4+ T-cell clones isolated from peripheral blood of a cured mucosal patient infected with *L. braziliensis* and a patient with subclinical infection with *L. donovani* proliferated and produced IFN-γ to many one- and two-dimensional immunoblotted parasite fractions. Likewise, Reed et al. (37) demonstrated that CD4+, IL-2-, and IFN-γ-secreting T-cell clones isolated from peripheral blood of a visceral patient with *L. chagasi* infection responded to 42- and 30-kDa proteins. These studies suggested that CD4+ IFN-γ-producing T-cells contribute to the resolution of human leishmaniasis. In studies of the distribution of cells in lesions of leishmaniasis, Modlin's group (23) demonstrated that CD4+ CD45RO+ memory cells predominated in both localized cutaneous and disseminated mucocutaneous lesions as well as in Montenegro skin reactions. Surprisingly, the percentage of cells containing IFN-γ mRNA as detected by in situ hybridization was similar. CD8+ T cells were present in both types of lesion but not in DH reactions elicited by *Leishmania* antigen, suggesting a role for these cells in the tissue destruction associated with lesions. We have shown that antigen-specific T-cell lines derived from PBMCs of patients with cutaneous or mucosal disease against *L. amazonensis* promastigote lysate or gp63 contained CD8+ as well as CD4+ T-cells. CD8+ cells coexpressing the $\alpha\beta$ T-cell receptor (TCR) or the $\gamma\delta$ TCR were found in T-cell lines against promastigote lysate (38). Together, this evidence indicates that CD8+ T-cells recognize parasite antigens and further suggests that these cells influence the host–parasite relationship.

A population of T-cells has been shown to express a TCR heterodimer distinct from that of the $\alpha\beta$-chain receptor. This new receptor was designated $\gamma\delta$ (39,40). It is expressed in a small population of thymocytes, dendritic epidermal cells, and intestinal intraepithelial cells (41–44). The role $\gamma\delta$ T-cells play in the induction and/or expression of immune responses to infectious organisms remains unknown, but the cells have been shown to possess many of the effector functions associated with $\alpha\beta$ T-cells, including cytotoxicity (45) and cytokine secretion (46). Increased numbers of these cells have been found in individuals with malignancies (47), immunodeficiencies (48), and a number of infectious diseases (49–51) including *Leishmania* (38,52).

Increasing evidence suggests their importance in the immune response to the intracellular pathogen *Mycobacterium* (53,54). In addition, $\gamma\delta$ T-cells have also been isolated from the autoimmune lesions of rheumatoid arthritis (55) and sarcoidosis (56). Therefore these cells may contribute to host defences against *Leishmania* or to the pathology associated with certain types of leishmaniasis.

In studies of leishmaniasis lesions, Modlin et al. (52) demonstrated that $\gamma\delta$ T-cells accumulated in localized cutaneous lesions but not in mucosal lesions or DH reactions to *Leishmania* antigens. These cells produced a factor that induced macrophage aggregation and proliferation and appeared to play a role in granuloma formation. The authors suggested that $\gamma\delta$ T-cells mediate the formation and organization of mature epithelioid granulomas and that their absence in mucosal disease may contribute to uncontrolled tissue destruction.

Studies conducted in our laboratory investigated $\gamma\delta$ T-cells found in peripheral blood of leishmaniasis patients (38). Initial observations revealed that patients with different forms of leishmaniasis (including cutaneous, mucosal, and visceral disease) all had elevated levels of circulating $\gamma\delta$ T-cells (Fig. 5.1). In most patients a significant percentage (25–50%) of these cells coexpressed the CD8 antigen, suggesting the potential for interaction with antigen in a class I restricted manner. In many instances, levels of circulating $\gamma\delta$ T-cells remained high in patients with cured or recently treated disease. Sequential blood samples obtained from four cutaneous patients, one prior to treatment and one 3–10 weeks after treatment, demonstrated that in three of the four patients $\gamma\delta$ T-cell percentages increased or remained elevated (>10%); only one patient showed a decrease in these cells. In other patients elevated levels of $\gamma\delta$ T-cells were noted up to 1 year after disease onset, but data regarding $\gamma\delta$ T-cell percentages in these patients prior to treatment was unavailable.

Antigen-specific T-cell lines generated from cutaneous or mucosal patients by stimulation with *L. amazonensis* lysate contained 25-60% $\gamma\delta$ T-cells, strongly suggesting that these cells are *Leishmania*-reactive (Fig. 5.2). In all seven individuals tested, the percentages of $\gamma\delta$ T-cells present in the T-cell lines increased dramatically (two- to eightfold) over that observed in the patients' PBMCs. All of these T-cell lines proliferated strongly to *L. amazonensis* lysate. It is evident that $\gamma\delta$ T-cells present in the peripheral blood of these patients have the potential to respond to *Leishmania* antigens. Thus the reported lack of $\gamma\delta$ T-cells in mucosal lesions may not be due to a failure in $\gamma\delta$ T-cell activation in leishmaniasis patients but may result from a trafficking phenomenon that prevents these cells from homing to the affected tissues.

In studies attempting to identify defined *Leishmania* antigen(s) that

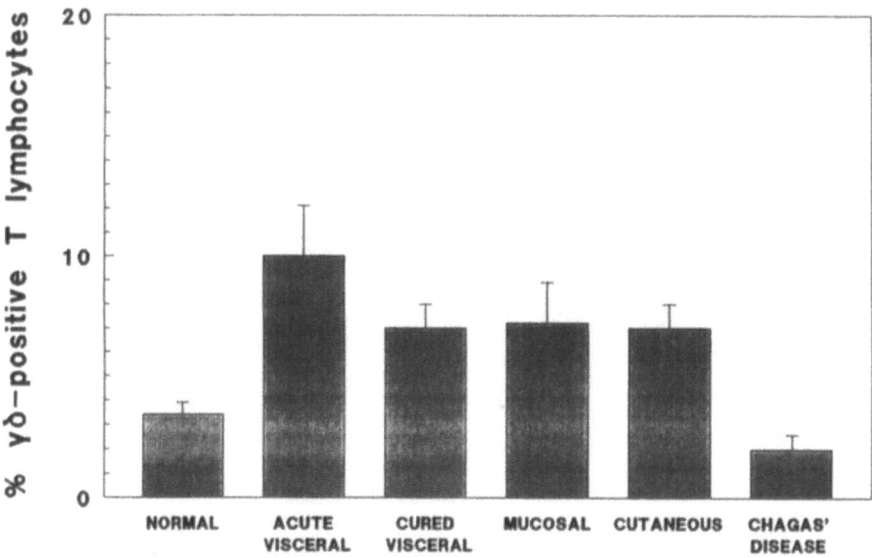

Figure 5.1. γδ T-cells in leishmaniasis patients. Resting PBMCs from normal individuals or patients with acute or cured visceral, mucosal, cutaneous leishmaniasis, or Chagas' disease were stained with TCRδ1 and analyzed by flow cytometry. Data are represented as the mean ± SEM. Statistical analysis was performed using a two-tailed Student's t-test. Patient groups were compared to normals ($n = 23$): acute visceral ($n = 8$), $p < 0.02$; postvisceral ($n = 7$), $p < 0.02$; mucosal ($n = 8$), $p < 0.05$; cutaneous ($n = 18$), $p < 0.002$; Chagas' disease ($n = 6$), not significant.

stimulate γδ T-cells, we generated T-cell lines against the dominant *L. amazonensis* surface antigens (gp63 and gp42) from PBMCs of two patients. Lines derived from these individuals against *L. amazonensis* lysate contained high percentages of γδ T-cells, in contrast lines stimulated with gp63 and gp42, did not contain γδ T-cells. This finding indicates that these two immunodominant antigens, which clearly stimulate strong αβ T-cell responses in leishmaniasis patients (57), do not activate γδ T-cells but that other antigens present in the *L. amazonensis* lysate stimulate these cells. Increasing evidence obtained from both human and murine systems demonstrates that a subset of γδ T-cells recognize heat shock proteins (HSPs) (58). The recognition of HSP provides one mechanism accounting for the proposed role of γδ T-cells in the elimination of autologous cells stressed by infection or other stimuli. Many pathogens, including *Leishmania* (59,60), can synthesize HSP in a threatening environment. It is possible that parasite-infected macrophages display altered self antigens that cross-react with parasite HSPs, or that parasite HSPs are pre-

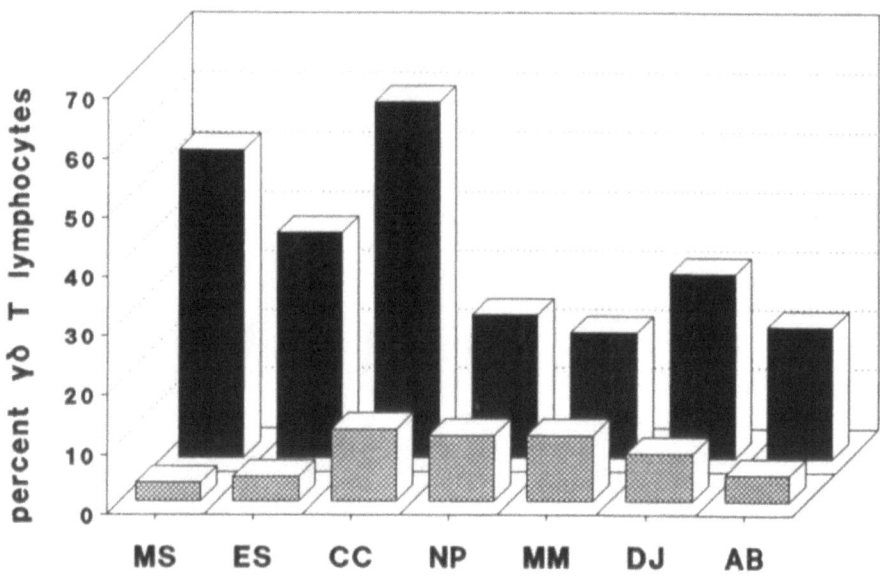

Figure 5.2. γδ T-cells in seven patients with PBMC (cross-hatched bars) versus antigen-stimulated cultures (solid bars). T-cell lines were generated by stimulating cutaneous or mucosal leishmaniasis patients' cells with *L. amazonensis* or *L. braziliensis* lysate.

sented by the macrophage in the conventional manner. Either scenario provides a signal that may activate γδ T-cells to lyse the infected cell. This mechanism may reduce parasite burdens but if uncontrolled could also cause considerable tissue damage. Other studies demonstrate that the specificity of γδ T-cells is not restricted to HSPs. These cells respond to mycobacterial proteins other than HSP (54) as well as to tetanus toxoid (46) in an MHC-restricted manner.

Leishmania Antigens that Stimulate Human T-Cells

Currently, there has been little success in the development of immunoprophylactics against *Leishmania*. In South America and the Middle East immunization has been tried with inactivated promastigotes or soluble parasite material. These attempts have been largely unsuccessful or difficult to evaluate. Yet murine studies clearly indicate that immunization is feasible. In light of this finding, emphasis in the study of *Leishmania* immunity has turned to the evaluation of purified *Leishmania* antigens in the hope of defining proteins or epitopes that can induce beneficial immunologic responses without inducing those that may be pathologic. Clearly, the identification of defined parasite antigens that selectively activate specific T-cell populations or induce T-cells to secrete certain cytokines would greatly contribute to our understanding of the host-parasite relationship and could yield potentially useful immunogens. Currently, our knowledge regarding the induction and expression of human T-cell responses to defined *Leishmania* antigens remains limited.

One antigen, gp63, is a cell surface protein of *Leishmania* that is highly conserved among species (61,62). In vitro, gp63 has been implicated in the ability of the parasite to invade macrophages (63); and in vivo studies showed that mice could be protected against cutaneous disease by immunization with gp63 incorporated into liposomes (64). In one human vaccination trial, lymphocytes from individuals immunized with killed *Leishmania* promastigotes proliferated in vitro to *L. amazonensis* gp63 (65). The study suggested that gp63 was an immunodominant antigen in the vaccine and induced, as well as elicited, T-cell responses. Button et al. (66) used recombinant *L. major* gp63 expressed in *E. coli* to derive antigen-specific T-cell lines and clones from a patient infected with *L. braziliensis*, thereby demonstrating the presence of conserved T-cell epitopes among Old and New World species of *Leishmania*.

Studies conducted in our laboratory evaluated cutaneous and mucosal patient responses to *L. amazonensis* native and recombinant gp63 (67). We found that native and recombinant gp63 induced strong in vitro proliferative responses in most patients tested (Table 5.1). Both the native and recombinant protein elicited the production of IFN-γ from patients' cells. In addition, T cell lines generated against native gp63 secreted IFN-γ, IL-2 but not IL-4 in response to promastigote lysate, native gp63 or recombinant gp63. This profile of cytokine secretion is characteristic of a Th1 type of immune response and considering the importance of IFN-γ in immune responses against *Leishmania*, suggests that gp63 is a potent and beneficial T cell immunogen. To further evaluate recombinant gp63, we immunized cells from normal donors in vitro with recombinant gp63. T cells induced in this way did not respond to subsequent stimulation with native gp63 or whole parasite lysate, although their response to the recombinant protein was high. This finding suggests that immunization with recombinant gp63 primes T cells which recognize epitopes conformationally distinct from those of the native protein; thus T cells induced by the recombinant protein may not be effectively elicited by challenge with live parasites.

In studies utilizing synthetic peptides of gp63 predicted by computer algorithms to contain T-cell epitopes, Jardim et al. (68) found that

Table 5.1 Patient responses to native versus recombinant gp63.

Antigens (in vitro)	Cutaneous patients				Mucosal patients			
	N		JM		DJ		AB	
	cpm (×10⁻³)	S.I.	cpm (×10⁻³)	S.I.	cpm (×10⁻³)	S.I.	cpm (×10⁻³)	S.I.
La	15.3 ± .2	11.0	40.3 ± 3.0	51.0	48.3 ± 12.3	11.0	83.2 ± 22.6	65.0
Nat. La gp63	5.0 ± 2.5	3.6	22.2 ± 5.5	28.0	31.7 ± 11.0	7.0	77.7 ± 5.3	60.0
Rec. La gp63	1.2 ± .5	0.9	12.7 ± 10.5	16.0	27.8 ± 21	6.0	29.0 ± 19.4	23.0
Rec. Lm gp63	N.D.	—	14.0 ± 8.9	18.0	N.D.	—	30.5 ± 11.0	23.0
No antigen	15.3 ± .8	1.0	.8 ± 3.0	1.0	4.5 ± .6	1.0	1.2 ± .5	1.0

PBMC were cultured for 5 days in 96-well flat-bottomed plates at 4×10 cells/well with *Leishmania* lysate (10 μg/ml); recombinant *La* or *Lm* gp63 (20 μg/ml), or without antigen. H-thymidine 1 μCi was added on day 5 for 18 hours.

three of seven peptides tested stimulated proliferation in several strains of mice, but only one peptide (PT3) protected mice against challenge with *L. major*. Studies conducted in our laboratory have shown that this peptide (PT3) does not elicit a proliferative response from leishmaniasis patient T-cells, but that an overlapping peptide (PT4) covering much of the same highly conserved region does. Current studies are emphasizing identification of T-cell epitopes (particularly those relevant in human responses) of gp63 and other antigens as well.

Other integral membrane proteins of *Leishmania*, most notably a 44 to 46-kDa protein of *L. amazonensis*, also have immunoprophylactic potential. Monoclonal antibodies raised against this protein passively protected mice from challenge infection (69). Immunization with gp46 and *C. parvum* also protected mice against *L. amazonensis* infection (70). This membrane glycoprotein elicited strong proliferative responses in PBMCs, from several cutaneous and mucosal patients, although the production of IFN-γ was not noted (57). We have used *Leishmania* patient sera to identify other antigens of *L. chagasi*. Selected antigens were purified and tested for effectiveness in stimulating patient PBMCs. Two glycoproteins, gp30 and gp42, stimulated potent proliferative responses in both cured visceral and mucosal patients (37). In addition, both antigens stimulated the production of IL-2 and IFN-γ by patient T-cells. Most interestingly, these two antigens appear to share T-cell epitopes. T-cell lines selected with gp30 or gp42 proliferated in response to either antigen as well as to *Leishmania* lysate. Among T-cell clones selected with *Leishmania* lysate, 13 responded to gp30, 11 to gp42, and 9 to both antigens. Epitope mapping is currently in progress. The identification of cross-reactive T-cell epitopes may provide an approach toward defining peptides that can induce or elicit appropriate T-cell responses.

Lipophosphoglycan (LPG) is a cell surface glycoconjugate comprised of repeating disaccharide units attached to a phosphatidylinositol lipid anchor by a phosphosaccharide core. LPG is expressed in all species of *Leishmania* and has been successfully used to vaccinate mice against *L. major* infection (71). In vitro assays suggests that LPG can inhibit oxidative burst activity in human monocytes as well as scavenge oxidative metabolites (72). Both mechanisms may contribute to the successful parasitism of monocytes by *Leishmania*.

T-lymphocyte responses to LPG have been reported in murine (73) and human (74) systems. The interpretation of these results has been complicated by the difficulty of obtaining LPG free of protein contaminants. Thus it has been difficult to determine whether the T cells are responding to protein or carbohydrate determinants (or both). The results obtained with the whole molecule in mouse experiments

Table 5.2 Reactivity of an LPG-specific line to components of LPG.

	M.LPG	
Antigen	x̄ cpm ± SD	S.I.
Lipophosphoglycan	7731 ± 1018	18.5
P-Disaccharide	752 ± 331	1.8
Core-phosphatidylinositol	6348 ± 661	15.2
Phosphoglycan	5822 ± 497	14.0
None	416 ± 42	1.0

T cells from an LPG-specific T-cell line were plated with irradiated autologous antigen presenting cells and LPG (20 μg/ml) or purified components of LPG (20 μg/ml). Cells and antigen were incubated for 3 days and pulsed with ^{3}H-thymidine for the final 6 hours of culture.

were presented as evidence for the induction of strong T-cell responses by carbohydrate, but firm evidence for this suggestion is lacking. One group demonstrated that LPG stimulated PBMCs from cutaneous leishmaniasis patients, but treatment of the molecule with proteinase K abolished its stimulatory activity (74a), suggesting that protein co-purifying with LPG was responsible for the T-cell stimulation. Work performed in our laboratory corroborated these studies (74b). We found that LPG elicited in vitro proliferative and in vivo DH responses in cutaneous, visceral, and mucosal patients. We generated an LPG-specific T-cell line and tested this line against various components of the LPG molecule, including the repeating disaccharide units, the core-phosphatidylinositol anchor portion, and phosphoglycan (Table 5.2). The T-cell line responded to the whole LPG molecule, to phosphoglycan, and to the core-anchor unit, demonstrating that the carbohydrate residues were not responsible for T-cell stimulation (74b). Still the immunogenicity of LPG cannot be dismissed. It may provide an important immunostimulatory effect for the protein it carries.

The characterization and delineation of the repertoire of T-cell populations that potentially respond to these antigens and the conditions required to induce the most beneficial effects is essential before immunization attempts with defined antigens can be considered in humans.

Importance of Cytokines in Leishmaniasis

The rapid expansion in molecular biology and its application to immunology has in recent years greatly increased our capacity to elucidate the cellular and molecular events involved in immune responses to pathogenic organisms. Several of the molecules that mediate cellular immune responses have been identified and cloned, and they are available in sufficient quantities to allow in vitro and in vivo studies to dissect their roles in immune function. Cytokines have particular relevance for parasitic infections for several reasons. The importance of T-cell responses in recovery from and resistance to many of these infections, such as leishmaniasis, is well documented. Additionally, T-cell responses may be deficient during certain parasitic infections. A list of cytokines involved in human or experimental leishmaniasis is shown in Table 5.3 and a model showing proposed cytokine-cellular interactions in leishmaniasis is shown in Figure 5.3.

Immune dysfunction appears to be directly related to parasite persistence and multiplication in VL. Lymphocytes from patients with acute disease fail to produce IL-2 or IFN-γ upon stimulation with *Leishmania* antigens, although these responses are not totally absent on mitogen stimulation (15). Upon recovery, *Leishmania*-induced cytokine production occurs, paralleling positive in vitro proliferative responses and positive DH responses. Immunomodulators are likely candidates to reverse such a transitory immunological incapacity. Such a therapeutic approach would thus be directed at restoring overall immune responsiveness, of critical importance in acute VL patients who often develop other infections. It illustrates the compelling rationale for considering cytokine therapy for parasitic infections, particularly for the protozoa that infect macrophages. Such therapy could be directed toward restoration of host immune responsiveness, as we have done successfully in animal models with a variety of cytokines (94–98), as well as toward limiting parasite proliferation. The importance of this approach is underscored by the lack of effective and nontoxic drugs for the treatment of leishmaniasis.

Immunotherapy directed at limiting parasite replication in host macrophages is a logical and particularly promising approach in leishmaniasis. The T-cell product, IFN-γ, activates macrophages, the primary host cells for *Leishmania*, to inhibit intracellular replication of the parasite. The anti-*Leishmania* effects of IFN-γ have been well documented in vitro (25,83). In vivo, significant reduction of parasite numbers were obtained in *L. chagasi*-infected mice treated with either supernatant from activated lymphocytes or recombinant IFN-γ encapsulated in liposomes (86). It was the first demonstration of successful therapy of a parasitic infection with cytokine. Although these animal studies achieved at best a 60% reduction in parasite numbers, this level of reduction could be sufficient to make a critical difference in clinical outcome. IFN-γ treatment was similiarly effective in mice infected with *L. donovani* (85). Further studies demonstrated that IFN-γ could synergize with pentavalent antimony, the drug of choice

Table 5.3 Cytokines demonstrated to have importance in leishmaniasis.

Cytokine	General/immunologic properties	Findings in leishmaniasis	Comments on immunotherapeutic potential
IL-1	Activates T cells. ↑ B cell proliferation and differentiation. ↑ Hematopoiesis in vivo. Induces APP.	Production ↓ *Leishmania*-infected MO in vitro (75). Normal production by PBMC from patients (17).	Presently difficult to handle clinically. May be useful as an adjuvant (76).
IL-2	↑ T & B cell growth. Activates MO. ↑ NK & LAK activity.	↓ Production by VL patients' lymphocytes during active disease (15).	Effective in CL (local administration) (77). Systemic use has major side effects. Likely candidate due to its importance in CMI.
IL-3	↑ Hematopoiesis in vivo and Eos activity and B-cell differentiation.	↑ In vitro MO leishmanicidal capacity. IL-3 in vivo ↑ disease progression in mice (78,79).	Needs careful evaluation because of possibility of worsening disease progression.
IL-4	↑ Activates T & B cell growth. Activates MO. ↑ IgE production.	High production by susceptible animals; anti-IL-4 Abs protect infected mice (32).	Blocking IL-4 may be effective.
M-CSF	↑ MO colonies. Activates MO.	↑ Human MO leishmanicidal activity (80).	Potential candidate. Needs testing.
G-CSF	↑ in vivo hematopoiesis. Activates MO and PMN granulocyte activity.		May be useful for correcting leukopenia of VL patients, and as MO activator. Similar to G-CSF.
GM-CSF	↑ In vivo hematopoiesis. Activates MO, PMN, Eos. Induces T-cell proliferation.	↑ Human MO leishmanicidal activity (26). In vivo ↑ disease progression in mice (81,82).	
IFN-γ	Activates MO. ↑ T & B cells. ↑ MHC Class-II antigens.	↑ MO leishmanicidal activity in vitro (25,83). In vivo is ineffective in mice with CL (84) and effective in VL (85,86). Synergizes the leishmanicidal effect in Sbᵛ (87).	Effective in CL (local administration) (88) and in VL (associated with antimony) patients (89).
TNF-α	Activates MO (synergizes with IFN-γ).	Elevated in acute VL patients (90,91), with rapid fall after cure (91). Administration protects infected mice (92,93).	Possible clinical use as immunomodulator may be limited by its dangerous interactions with IFN-γ.

APP = acute phase protein; Eos = eosinophils; PMN = neutrophils; MO = macrophages; Sbᵛ = antimony; ↑ = stimulates; ↓ = decreases.

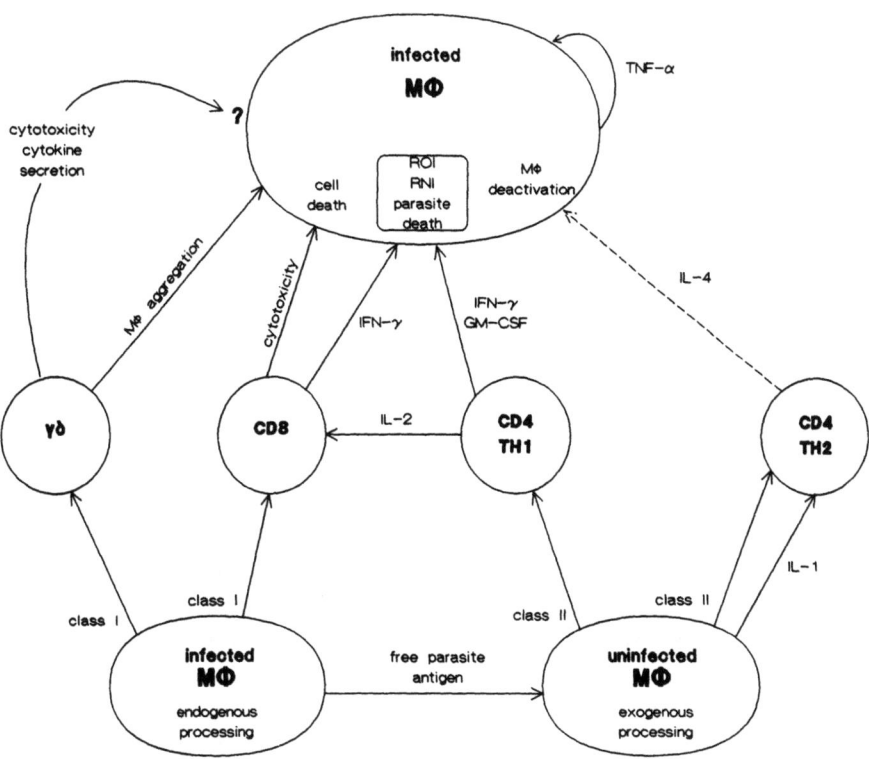

Figure 5.3. T-cell-cytokine interactions postulated to play a role in the immune response to *Leishmania* infection. Solid arrows indicate activating pathways; broken arrows indicate deactivating pathways. ROI = reactive oxygen intermediates; RNI = reactive nitrogen intermediates.

for all forms of leishmaniasis, for the inhibition of *Leishmania* (87). It indicated that a combination of chemo- and immunotherapy could also have potential in a more effective treatment approach.

Clinical trials have supported this concept. IFN-γ was used in conjunction with antimony for the treatment of refractory visceral leishmaniasis (89). In this study, patients who had not responded to treatment with antimony alone were successfully treated with the combination approach. Successful treatment was assessed by increases in leukocyte counts, spleen size regression, and other clinical parameters. Thus the inclusion of IFN-γ in the chemotherapeutic regimen reversed unresponsiveness to the antimony alone. Traditionally, patients failing antimony therapy receive amphotericin B, a drug that is toxic and difficult to manage. Our demonstration of the effectiveness of antimony and IFN-γ opens new possibilities for the treatment of visceral leishmaniasis. Other investigators have reported clinical treatment of CL with IFN-γ (88). These studies demonstrate the practical clinical application of cytokine therapy to a parasitic infection and support the importance of T-cell-derived cytokines for the control of *Leishmania*.

Other cytokines have been used in clinical or experimental *Leishmania* infection, including the colony-stimulating factors GM-CSF or G-CSF. Both of these cytokines have the ability to stimulate hematopoiesis in vivo, both have been applied clinically, and each has received U.S. Food and Drug administration (FDA) approval. The ability to stimulate bone marrow cells, particularly granulocytes, may be important for correcting the leukopenia and protecting VL patients from bacterial infections, which can be a cause of death in these patients. GM-CSF may be particularly promising. In addition to stimulating hematopoiesis, this cytokine can also increase T-cell function (94), induce macrophage cytotoxicity (99), and stimulate macrophage inhibition of *Leishmania* (26). This multipotential cytokine may therefore have effects on several aspects of the treatment of infectious disease. Another colony-stimulating factor, M-CSF, has also been shown to induce leishmanicidal activity in human macrophages (J. Ho, personal communication). All of these cytokines deserve careful consideration for application in leishmaniasis as well as other infections. As a note of caution, however, there have been reports of exacerbation of *Leishmania* infections in mice treated with GM-CSF (81,82) or IL-3 (78,79). It was suggested that this negative effect may have been the result of an increase in host cell macrophages resulting from the treatments, but this hypothesis remains to be proved. Although the mouse models for human leishmaniasis are in general dissimilar, these studies raise possibilities that should be considered when embarking on new immunotherapeutic strategies.

Another cytokine that may be useful in *Leishmania* infections is IL-2. Some inhibition of *L. donovani* was achieved in mice using IL-2, and local therapy of CL lesions with IL-2 was reported to be successful in humans (77). Systemic administration of IL-2 has been shown to increase levels of IFN-γ (100), but toxicity is an important consideration in the therapeutic use of this cytokine.

Conclusions

Several aspects of *Leishmania* infections point to the importance of T-cells in these diseases. In addition, *Leishmania* infections, both clinical and experimental, have contributed significantly to understanding basic T-cell function. Recovery from and immunity to *Leishmania* infections relies on cellular immune responses. Thus parasite-induced lymphocyte proliferation responses, T-cell cytokine production, and DH responses are all negative during acute VL when specific antibody production is high. T-cell responses are positive on recovery, persisting for several years. Healing CL is characterized by strong T-cell responses and diffuse CL by weak T-cell responses, high levels of specific antibody, and an inability to control the infection. Quiescent *Leishmania* infections can be reactivated by immunosuppressive treatment or by HIV infection, indicating the persistent nature of the parasite and the importance of intact T-cell responses in preventing disease. On the other side is destructive ML, the pathology of which is associated with a hyperactive T-cell response.

Leishmania infections are also providing a valuable model for understanding T-cell function. Mouse infections have indicated that activation of different T-cell pathways may determine whether the outcome will be healing or death. Mice may be protected from fatal infections by partially ablating T-cell responses. Passive transfer of T-cells can lead to either healing or exacerbation of infections, depending on T-cell phenotype and antigen specificity. In humans, the presence of high levels of circulating $\gamma\delta$ T-cells provides a unique clinical situation and an excellent opportunity for elucidating the function of these cells.

Advances in the identification and cloning of *Leishmania* antigens have provided tools to dissect specific T-cell responses. In a system in which antigens may induce both exacerbative and protective responses, the dissection of T-cell responses to defined epitopes may lead to the development of a vaccine that induces protective responses without the danger of inducing destructive ones. Identification of T-cell-stimulating antigens and epitopes within these antigens thus remains a challenge for future studies. Such developments parallel those being made in the development of improved delivery systems and adjuvants for inducing strong cellular responses in humans. In this regard, the potential of cytokines as adjuvants remains to be explored.

Further developments in the areas of cytokine therapy will undoubtedly be of importance to leishmaniasis. Initial success points to the potential for using T-cell products directly for treating a disease in which T-cell function is deficient. Cytokine therapy promises to be of explosive importance for the treatment of many types of disease. These advances have been made to treat patients in developed countries, but they will also benefit patients with diseases in which the same degree of intense research effort has not been done.

References

1. Peters W, Killick-Kendrick R (eds): The Leishmaniasis in Biology and Medicine. Vol. II. Clinical Aspects and Control. Academic Press, Orlando, FL, 1987.

2. Chang KP, Bray RS (eds): Human Parasitic Diseases. Vol. 1. Leishmaniasis. Elsevier Science Publishers, New York, 1985.
3. Manson-Bahr PEC: Active immunization in leishmaniasis. In Garnham PCC, Pierce AE, Ratt I (eds): Immunity to Protozoa. Blackwell Scientific, Oxford, 1963, p. 246.
4. Montenegro J: A cutis-reaccao na leishmaniose. Ann Fac Med Univ Sao Paulo 1926;1:323.
5. Reed SG, Badaro R, Masur H, et al: Selection of a skin test antigen for American visceral leishmaniasis. Am J Trop Med Hyg 1986;35:79.
6. Badaro R, Pedral-Sampaio D, Johnson WD Jr, Reed SG: Evaluation of the stability of a soluble intradermal skin test antigen preparation in American visceral leishmaniasis. Trans R Soc Trop Med Hyg 1990;84:226.
7. Castes M, Agnelli A, Verde O, Rondon AJ: Characterization of the cellular immune response in American cutaneous leishmaniasis. Clin Immunol Immunopathol 1983;27:176.
8. Carvalho EM, Johnson WD, Barreto E, et al: Cell mediated immunity in American cutaneous and mucosal leishmaniasis. J Immunol 1985; 135:4144.
9. Manson-Bahr PEC: The leishmanin test and immunity in Kala-azar. E Afr Med J 1961;38:165.
10. Manson-Bahr PEC, Heisch RB, Garnham PCC: Studies in leishmaniasis in East Africa: the Montenegro test in Kala-azar in Kenya. Trans R Soc Trop Med Hyg 1959;53:380.
11. Andrade TM, Teixeira R, Andrade JAF, et al: Estudo da hipersensibilidade de tipo retardado na leishmaniose visceral. Rev Inst Med Trop Sao Paulo 1982;24:298.
12. Badaro R, Jones TC, Carvalho EM, et al: New perspectives on a subclinical form of visceral leishmaniasis. J Infect Dis 1986;154:1003.
13. Mayrink W, Williams P, Costa CA, et al: An experimental vaccine against American dermal leishmaniasis: experience in the state of Espirito Santo. Ann Trop Med Parasitol 1985;79:259.
14. Carvalho EM, Teixeira RS, Johnson Jr WD: Cell mediated immunity in American visceral leishmaniasis. Infect Immun 1981;48:409.
15. Carvalho EM, Badaro R, Reed SG, et al: Absence of gamma interferon and interleukin 2 production during active visceral leishmaniasis. J Clin Invest 1985;76:2066.
16. Sacks DL, Lata Lal S, Shrivastava SN, et al: An analysis of T cell responsiveness in Indian kala-azar. J Immunol 1987;138:908.
17. Cillari E, Liew FY, Lo Campo P, et al: Suppression of IL-2 production by cryopreserved peripheral blood mononuclear cells from patients with active visceral leishmaniasis in Sicily. J Immunol 1988;140:2721.
18. Wyler DJ: Circulating factor from a Kala-azar patient suppresses in vitro anti-leishmanial T cell proliferation. Trans R Soc Trop Med Hyg 1982;76:304.
19. Barral A, Carvalho EM, Badaro R, Barral-Netto M: Suppression of lymphocyte proliferative responses by sera from patients with American visceral leishmaniasis. Am J Trop Med Hyg 1986;35:735.
20. Carvalho EM, Bacellar O, Barral A, et al: Antigen-specific immuno-

suppression in visceral leishmaniasis is T cell mediated. J Clin Invest 1989;83:860.
21. Ridley DS, Ridley MJ: The evolution of the lesion in cutaneous leishmaniasis. Int J Dermatol 1983;18:50.
22. Barral A, Jesus A, Almeida RP, et al: Evaluation of T cell subsets in the lesion infiltrates of human cutaneous and mucocutaneous leishmaniasis. Parasite Immunol 1987;9:487.
23. Pirmez C, Cooper C, Paes-Oliveira M, et al: Immunologic responsiveness in American cutaneous leishmaniasis lesions. J Immunol 1990; 145:3100.
24. Nathan CF, Murray HW, Wiebe ME, Rubin BY: Identification of interferon gamma as the lymphokine that activates human macrophages oxidative metabolism and antimicrobial activity. J Exp Med 1983;158:670.
25. Murray HW, Rubin BY, Rothermel CD: Killing of intracellular *Leishmania donovani* by lymphokine-stimulated human mononuclear phagocyte: evidence that interferon gamma is the activating lymphokine. J Clin Invest 1983;72:1506.
26. Weiser WY, Van Neil A, Clark SC, et al: Recombinant human granulocyte/macrophage colony-stimulating factor activates intracellular killing of *Leishmania donovani* by human monocyte-derived macrophages. J Exp Med 1987;166:1436.
27. Corradin SB, Mauel J: Phagocytosis of *Leishmania* enhances macrophage activation by IFN-γ and lipopolysaccharide. J Immunol 1991; 146:279.
28. Green SJ, Meltzer MS, Hibbs JB, Nacy CA: Activated macrophage destroy intracellular *Leishmania major* amastigotes by an L-arginine-dependent killing mechanism. J Immunol 1990;144:278.
29. Mosmann TR, Cherwinski H, Bond MW, et al: Two types of murine helper T cell clone. I. Definition according to profiles of lymphokine activities and secreted proteins. J Immunol 1986;136:2348.
30. Cherwinski HM, Schumacher JH, Brown KD, Mosmann TR: Two types of mouse helper T cell clone. III. Further differences in lymphokine synthesis between Th1 and Th2 clones revealed by RNA hybridization, functionally monospecific bioassays and monoclonal antibodies. J Exp Med 1987;166:1229.
31. Scott P, Natovitz P, Coffman R, Pearce E, Sher. A. Immunoregulation of cutaneous leishmaniasis: T cell lines that transfer protective immunity or exacerbation belong to different T-helper subsets and respond to distinct parasite antigens. J Exp Med 1988;168:1675.
32. Heinzel FP, Sadick MD, Holaday BJ, et al: Reciprocal expression of interferon gamma or interleukin 4 during the resolution or progression of murine leishmaniasis: evidence for expansion of distinct helper T cell subsets. J Exp Med 1989;169:59.
33. De Libero G, Kaufmann SHE: Antigen-specific Lyt2$^+$ cytolytic T lymphocytes from mice infected with the intracellular bacterium Listeria monocytogenes. J Immunol 1986;137:2688–2694.
34. Weiss WR, Sedegah M, Beaudoin RL, et al: CD8$^+$ T cells (cytotoxic/

suppressors) are required for protection in mice immunized with malaria sporozoites. Proc Natl Acad Sci USA 1988;85:573–576.
35. Melby PC, Neva FA, Sacks DL: Profile of human T cell response to leishmanial antigens: analysis by immunoblotting. J Clin Invest 1989; 83:1868–1875.
36. Melby PC, Sacks DL: Identification of antigens recognized by T cells in human leishmaniasis: analysis of T cell clones by immunoblotting. Infect Immun 1989;57:2971.
37. Reed SG, Carvalho EM, Sherbert CH, et al: In vitro response to *Leishmania* antigens by lymphocytes from patients with leishmaniasis or Chagas' disease. J Clin Invest 1990;85:690.
38. Russo DM, Barral-Netto M, Armitage RJ, et al: Identification of circulating γδ T cells in leishmaniasis patients. Submitted.
39. Brenner MB, McLean J, Dialynas DP, et al: Identification of a putative second T-cell receptor. Nature 1986;322:145.
40. Bank I, DePinho RA, Brenner MB, et al: A functional T3 molecule associated with a novel heterodimer on the surface of immature human thymocytes. Nature 1986;322:79.
41. Brenner MB, Strominger JL, Krangel MS: The gamma-delta T cell receptor. Adv Immunol 1988;43:133.
42. Lew AM, Pardoll DM, Maloy WL, et al: Characterization of T cell receptor gamma chain expression in a subset of murine thymocytes. Science 1986;334:1401.
43. Stingl G, Gunter KT, Tschachler E, et al: Thy-1+ dendritic epidermal cells belong to the T cell lineage. Proc Natl Acad Sci USA 1987;84:2430.
44. Goodman T, Lefrancois L: Expression of the gamma-delta T cell receptor on intestinal CD8+ intraepithelial lymphocytes. Nature 1988;333:855.
45. Palliard X, Yssel H, Basnnchard D, et al: Antigen specific and MHC nonrestricted cytotoxicity of T cell receptor αβ and γδ human T cell clones isolated in IL-4. J Immunol 1989;143:452.
46. Kozbor D, Trinchieri G, Monos, DS, et al: Human TCR-γ+/δ+, CD8+ T lymphocytes recognize tetanus toxoid in an MHC-restricted fashion. J Exp Med (1989);169:1847.
47. Seki S, Abo T, Masuda, T, et al: Identification of activated T cell receptor γδ lymphocytes in the liver of tumor-bearing hosts. J Clin Invest 1990;86:409.
48. Margolick JB, Scott, ER, Odaka N, Saah. A: Flow cytometric analysis of γδ T cells and natural killer cells in HIV-1 infection. Clin Immunol Immunopathol 1991;58:126.
49. Haas W, Kaufmann S, Martinez-A C: The development and function of γδ T cells. In Gallagher R (ed): Immunology Today. Vol. 11. Elsevier Trends Journals, Cambridge, 1990, P, 340.
50. Roussilhon C, Agrapart M, Ballet JJ, Bensussan A: T lymphocytes bearing the γδ T cell receptor in patients with acute Plasmodium falciparum malaria. J Infect Dis 1990;162:283.
51. Ho M, Webster HK, Tongtawe P, et al: Increased γδ T cells in acute Plasmodium falciparum malaria. Immunol Lett 1990;25:139.

52. Modlin RL, Pirmez C, Hofman F, et al: Lymphocytes bearing antigen specific $\gamma\delta$ T-cell receptors accumulate in human infectious disease lesions. Nature 1989;339:544.
53. Haregewoin A, Soman G, Hom RC, Finberg RW: Human $\gamma\delta^+$ T cells respond to mycobacterial heat-shock protein. Nature 1989;340:309.
54. Kabelitz D, Bender A, Schondelmaier S, et al: A large fraction of human peripheral blood $\gamma\delta$ T cells is activated by Mycobacterium tuberculosis but not by its 65 kd heat shock protein J Exp Med 1990;171:667.
55. Holoshitz J, Koning F, Coligan J, et al: Isolation of CD4$^-$ CD8$^+$ mycobacteria-reactive T lymphocyte clones from rheumatoid arthritis synovial fluid. Nature 1989;339:226.
56. Balbi B, Moller DR, Kirby M, et al: Increased numbers of T lymphocytes with $\gamma\delta$-positive antigen receptors in a subgroup of individuals with pulmonary sarcoidosis. J Clin Invest 1990;85:1353.
57. Burns Jr JM, Scott JM, Russo DM, et al: Characterization of a membrane antigen of *Leishmania amazonensis* which stimulates human immune responses. J Immunol 1990;146:742.
58. Born W, Hafpp M, Dallas A, et al: Recognition of heat shock proteins and $\gamma\delta$ cell function. Immunol Today 1990;11(2):40.
59. Searle S, Campos, AJ, Coulson RM, et al: A family of heat shock protein 70-related genes are expressed in the promastigotes of *Leishmania major*. Nucleic Acids Res 1989;17:5081.
60. MacFarlane J, Blaxter ML, Bishop RP, et al: Identification and characterization of a *Leishmania donovani* antigen belonging to the 70-kDa heat-shock protein family. Eur J Biochem 1990;190:377.
61. Colmer-Gould V, Quintao, LG, Keithly J, Nogueira N: A common major surface antigen on amastigotes and promastigotes of *Leishmania* species. J Exp Med 1985; 162:902.
62. Reed SG, Badaro R, Lloyd RMC: Identification of specific and cross-reactive antigens of *Leishmania donovani chagasi* by human infection sera. J Immunol 1987; 138:1596.
63. Russell DG, Wilheim H: The involvement of the major surface glycoprotein (gp63) of *Leishmania* promastigotes in attachment to macrophages. J Immunol 1986;138:1596.
64. Russell DG, Alexander J: Effective immunization against cutaneous leishmaniasis with defined membrane antigens reconstituted into liposomes. J Immunol 1988;140:1274.
65. Nascimento E, Mayrink W, da Costa CA, et al: Vaccination of humans against cutaneous leishmaniasis: cellular and humoral immune responses. Infect Immun 1990;58:2198.
66. Button LL, Reiner NE, McMaster WR. Modification of *Leishmania* gp63 genes by the polymerase chain reaction for expression of nonfusion protein at high levels in *Escherichia coli*: application to mapping protective T cell determinants. Mol. Biochem. Parasitol. 1991;44:213.
67. Russo DM, Burns JM, Carvalho E, et al: Human T cell responses to gp63, a surface antigen of *Leishmania*. J. Immunol. 1991;147:3575.
68. Jardim A, Alexander J, Teh HS, et al: Immunoprotective *Leishmania major* synthetic T cell epitopes. J Exp, Med (1990). 172:645.

69. Anderson S, David JR, McMahon-Pratt D: In vivo protection against *Leishmania mexicana* mediated by monoclonal antibodies. J Immunol 1983;131:1616.
70. Champsi J, McMahon-Pratt D: Membrane glycoprotein M-2 protects against *Leishmania amazonensis* infection. Infect Immun 1988; 52:3272.
71. Handman E, Mitchell GF: Immunization with *Leishmania* receptor for macrophages protects mice against cutaneous leishmaniasis. Proc Nat Acad Sci USA 1985;82:5910.
72. McNeely TB, Turco SJ: Immunization with *Leishmania* receptor for macrophages protects mice against cutaneous leishmaniasis. J Immunol 1990;144:2745.
73. Moll H, Mitchell GF, McConville MJ, Handman E: Evidence for T cell recognition in mice of a purified lipophosphoglycan from *Leishmania major*. Infect Immun 1989;57:3349.
74a. Mendonca SCF, Russel DG, Coutinho SG: Analysis of the human T cell responsiveness to purified antigens of *Leishmania*: Lipophosphoglycan (LPG) and glycoprotein 63 (gp 63). Clin. Exp. Immonol 1991;83:472.
74b. Russo DM, Turco SJ, Burns JM Jr. and Reed SG. 1992. Stimulation of human T lymphocytcs by *Leishmania* Lipophosphoglycan-associated proteins. J. Immunol 148:202.
75. Crawford GD, Wyler DJ, Dinarello CA: Parasite-monocyte interactions in human leishmaniasis: production of interleukin-1 in vitro. J Infect Dis 1985;152:315.
76. Staruch MJ, Wood DD: The adjuvanticity of interleukin-1 in vivo. J Immunol 1983;130:2191.
77. Akuffo H, Kaplan G, Kiessling R, et al: Administration of recombinant interleukin-2 reduces the local parasite load of patients with disseminated cutaneous leishmaniasis. J Infect Dis 1990;161:775.
78. Feng ZY, Louis J, Kindler VT, et al: Aggravation of experimental cutaneous leishmaniasis in mice by administration of interleukin 3. Eur J Immunol 1988;18:1245.
79. Lelchuk R, Graveley, R, Liew FY: Susceptibility to murine cutaneous leishmaniasis correlates with the capacity to generate interleukin 3 in response to *Leishmania* antigen in vitro. Cell Immunol 1988;111:66.
80. Ho JL, Reed SG, Wick EA, Giordano M: Granulocyte-macrophage and macrophage colony stimulating factors activate intramacrophage killing of *Leishmania mexicana amazonensis*. J Infect Dis 1990;162:224.
81. Griel J, Bodendorfer B, Rollinghoff M, Solbach W: Application of recombinant granulocyte-macrophage colony stimulating factor has a detrimental effect in experimental murine leishmaniasis. Eur J Immunol 1988;18:1525.
82. Corcoran LM, Metcalf D, Edwards SJ, Handman E: GM-CSF produced by recombinant vaccinia virus or in GM-CSF transgenic mice has no effect in vivo on murine cutaneous leishmaniasis. J Parasitol 1988; 74:763.
83. Hoover DL, Nacy CA, Meltzer MS: Human monocyte activation for cytotoxicity against intracellular L. donovani amastigotes induction of microbicidal activity by interferon gamma. Cell Immunol 1985;94:500.

84. Sadick MD, Heinzel FP, Holaday BJ, et al: Cure of murine leishmaniasis with anti-interleukin-4 monoclonal antibody: evidence for a T cell-dependent interferon-gamma independent mechanism. J Exp Med 1990;171:115.
85. Murray, HW, Stein J, Welte K: Experimental visceral leishmaniasis: production of interleukin-2 and gamma interferon, tissue immune reactions, and response to treatment with interleukin-2 and gamma interferon. J Immunol 1987;138:2290.
86. Reed SG, Barral-Netto M, Inverso JA: Treatment of experimental visceral leishmaniasis with lymphokine encapsulated in liposomes. J Immunol 1984;132:3116.
87. Murray HW, Berman JD, Wright SD: Immunochemotherapy for intracellular *Leishmania donovani* infection: gamma interferon plus pentavalent antimony. J Infect Dis 1988;157:973.
88. Harms G, Chehadi AK, Racz P, et al: Effects of intradermal gamma-interferon in cutaneous leishmaniasis. Lancet 1989;10:1287.
89. Badaro R, Falcoff E, Badaro FS, et al: Treatment of visceral leishmaniasis with pentavalent antimony and interferon gamma. N Engl J Med 1990;322:16.
90. Scuderi P, Lam KS, Ryan KJ, et al: Raised serum levels of tumor necrosis factor in parasitic infections. Lancet 1986;2:1364.
91. Barral-Netto M, Badaro R, Barral A, et al: Tumor necrosis factor Cachectin in human visceral leishmaniasis. J Infect Dis (in press).
92. Titus RG, Sherry B, Cerami A: Tumor necrosis factor plays a protective role in experimental murine cutaneous leishmaniasis. J Exp Med 1989;6:2097.
93. Liew FY, Parkinson C, Millott S, et al: Tumor necrosis factor (TNF alpha) in leishmaniasis. I. TNF alpha mediates host protection against cutaneous leishmaniasis. Immunobiology 1990;69:570.
94. Reed SG, Grabstein KH, Pihl D, Morrisey PJ: Recombinant granulocyte-macrophage colony-stimulating factor restores deficient immune responses in mice with chronic *Trypanosoma cruzi* infections. J Immunol 1990;145:1564.
95. Tarleton RL, Kuhn RE: Restoration of in vitro immune responses of spleen cells from mice infected with *Trypanosoma cruzi* by supernatants containing interleukin 2. J Immunol 1984;133:1570.
96. Reed SG, Inverso JA, Roters SB: Suppressed antibody responses to sheep erythrocytes in mice with chronic *Trypanosoma cruzi* infections are restored with interleukin-2. J Immunol 1984;133:3333.
97. Reed SG, Pihl DL, Grabstein KH: Immune deficiency in chronic *Trypanosoma cruzi* infection: recombinant interleukin-1 restores Th function for antibody production. J Immunol 1989;142:2067.
98. DeTitto EH, Catterall JR, Remington JS: Activity of recombinant tumor necrosis factor on *Toxoplasma gondii* and *Trypanosoma cruzi*. J Immunol 1985;157:1342.
99. Grabstein KH, Urdal DL, Tushinski RJ, et al: Induction of macrophage tumorcidal activity by granulocyte-macrophage colony-stimulating factor. Science 1986;232:506.
100. Lotze MT, Matory YL, Ettinghansen SE, et al: In vivo administration of

purified human interleukin-2. II. Half-life immunologic effects, and expansion of peripheral lymphoid cells in vivo with recombinant IL-2. J Immunol 1985;135:2865.

6

Evolution of Prosobranch Snails Transmitting Asian *Schistosoma*; Coevolution with Schistosoma: A Review

George M. Davis

More than a decade has passed since I wrote "Snail Hosts of Asian *Schistosoma* Infecting Man: Evolution and Coevolution" (1). Considerable progress has been made during this time in understanding the patterns and processes of evolution of triculine snails throughout Asia and the pattern of coevolution with schistosomes. This progress has been made possible through research initiatives in China and India, particularly as a result of initiating work in Yunnan, China in 1983.

A wealth of new morphological data has been obtained from numerous populations, allowing for refinements of triculine phylogenies previously given. A robust phylogeny of the Triculinae was essential to understand the coevolved relationships among some triculines and species of *Schistosoma*. Several nominal genera and species described long ago were searched for and located in India and in Yunnan and Hunan Provinces in China. Some of these taxa were especially critical for resolving long outstanding problems involving nomenclature and testing phylogenies based on the data previously assembled. As examples: (1) What was *Tricula* on which the subfamily Triculinae is based? (2) What was *Lithoglyphopsis*, the genus used to describe *Lythoglyphopsis aperta* Temcharoen, 1971 (now classified as *Neotricula aperta*), the snail host of *Schistosoma mekongi*? (3) What were the genera *Fenouilia* and *Delavaya* from Yunnan, China? Previous workers (2–4) considered these odd-looking (shells) genera to belong to a family unrelated to the Triculinae, whereas I (5, pp. 112–114) argued that they were indeed Triculinae. Indeed, my hypothesis was then, as it is today, that all freshwater rissoacean prosobranch snails of southern China and Southeast Asia [that were clearly not Stenothyridae or Bithyniidae (easily identifiable families)] were pomatiopsid—either Pomatiopsinae or Triculinae. Detailed anatomical studies of these taxa were essential for better understanding the phylogeny of the Triculinae.

In order to answer the above questions, it was necessary to locate the type species on which each genus was based. Of what importance is a name the reader might ask? Of what relevance today is nomenclature? Considerable! A name is a key to a unique encyclopedia of genetic information written over a considerable period of time. One name can involve several volumes containing the detailed genetic information relating to the interaction of a group of species of schistosomes and an evolving snail lineage. Careless use of a name may be totally misleading, inferring patterns and processes of snail-schistosome coevolution that have no bases in fact. For example, to say that hydrobiid snails are involved in the transmission of *Schistosoma japonicum* is to make an assertion that not only is wrong but would lead, if believed, to erroneous conclusions about the tightly woven genetic interaction between the *Schistosoma japonicum* species complex and a particular lineage of snails over 12 million years or more in a discrete region of space—space in which there is no evidence for the presence of the Hydrobiidae.

I previously established (5) that there are two highly divergent clades (lineages, subfamilies) of prosobranch snails transmitting species of *Schistosoma* that infect man; both were classified as Pomatiopsidae. *Oncomelania* of the subfamily Pomatiopsinae transmits *Schistosoma japonicum*; *Tricula* and closely related genera (i.e., *Tricula sensu lato*) of the subfamily Triculinae transmit other species of what Davis and Greer (16) described as members of the *S. japonicum* species complex. The central purpose of this chapter is to provide a synthesis of data allowing for considerable advances in knowledge involving the Triculinae.

A number of hypotheses established a decade ago (Davis 1,5,7) have been continuously challenged. They are reexamined here along with additional hypotheses made later (8,9).

1. *Schistosoma* and the relevant snail faunas evolved in Gondwanaland prior to the breakup of Pangaea (Fig. 6.1).
2. Coevolution of *Schistosoma* and Mesozoic snails began in the Gondwanian regions now known as Africa, India, and South America; the Asian schistosomes and relevant snails (both prosobranch and pulmonate) were introduced to the mainland of Asia via the Indian Plate, specifically through the Brahmaputra gap into northern Burma and Yunnan, China (5); *Schistosoma* and *Biomphalaria* were introduced to South America by the South American plate (Figs. 6.1 and 6.2).
3. Coevolution of *Schistosoma* and snail hosts involved reciprocal selective pressures affecting the course of evolution of both parasite and snail.
4. Specificity of the snail–parasite interaction, involving relatively few genes, increased exponentially through time with the creation of new ecological space due to the Himalayan orogeny. The rate of increasing specificity was dictated primarily by the rate of evolution of snail groups (i.e., speciation and cladogenesis).
5. Relevant snail faunas evolved more rapidly than *Schistosoma*.
6. Through time, *Schistosoma* has been restricted to snails of comparatively generalized phenotype and, by implication, generalized genotype.
7. The direction of evolution is down evolving river systems from those regions in Tibet and Yunnan, China where the major rivers (Yangtze, Mekong, Red, Salween, and Irrawaddy) are 100 miles or less apart. The beginning of the explosive stage of speciation and cladogenesis was 12 ± 4 million years ago.
8. Increasing morphological specializations/innovations were to be expected as one progressed downstream.
9. Taxa representing grades of morphology pertaining to the three tribes of Triculinae (5,10,20) (Fig. 6.3) had evolved in southwestern China (and perhaps northern Burma) during the early Pleistocene (or late Pliocene) prior to the period of lake level lowering in Yunnan, China and the disruption of cross-communications between the Yangtze and Mekong River drainages. Schistosome lineages associated with evolving associated triculine lineages would thus have considerable co-evolved divergences and specificities in this region at this period.

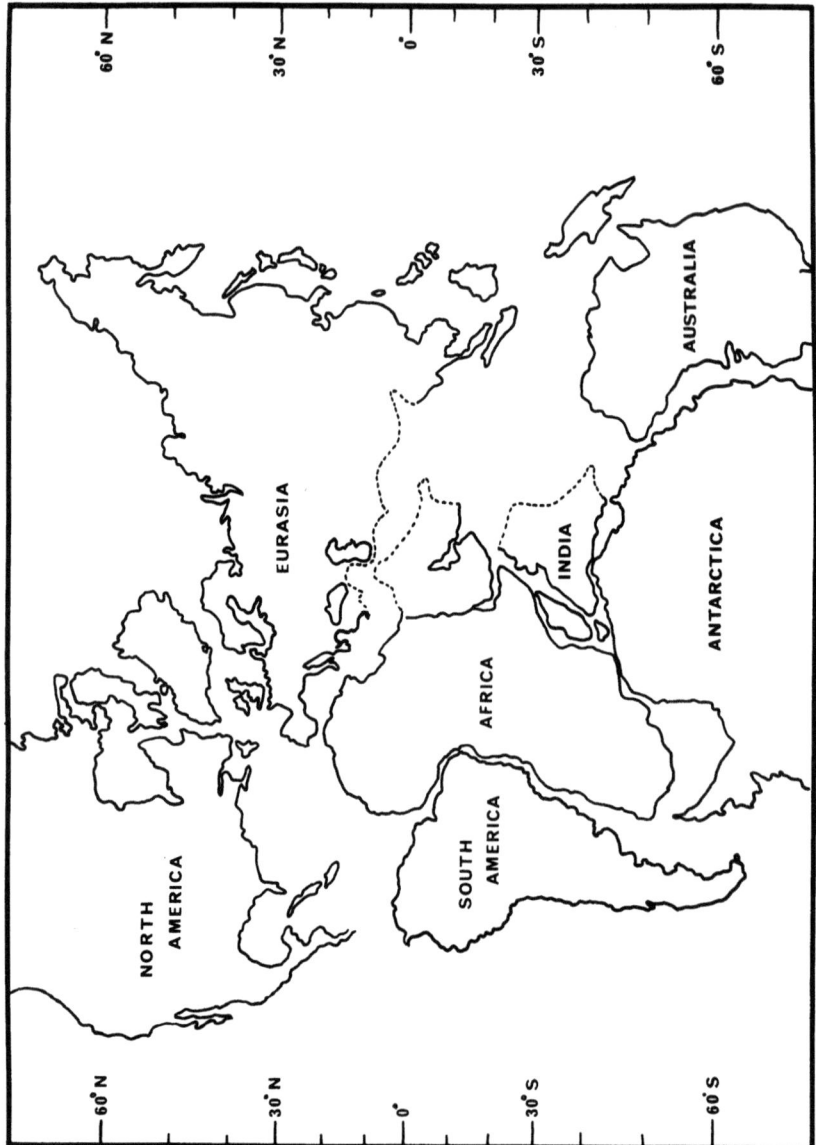

Figure 6.1. Gondwanaland in the Middle Cretaceous (>100 million years ago). Precursors of pomatiopsid and *Schistosoma* taxa are hypothesized to exist on the Indian and African Plates. Pomatiopsid taxa, given their current distribution, existed additionally on the South American and Australian and probably the Antarctic Plate. Adapted from Raven & Axelrod (62) and Davis (5).

6. Evolution of Prosobranch Snails Transmitting Asian *Schistosoma* 149

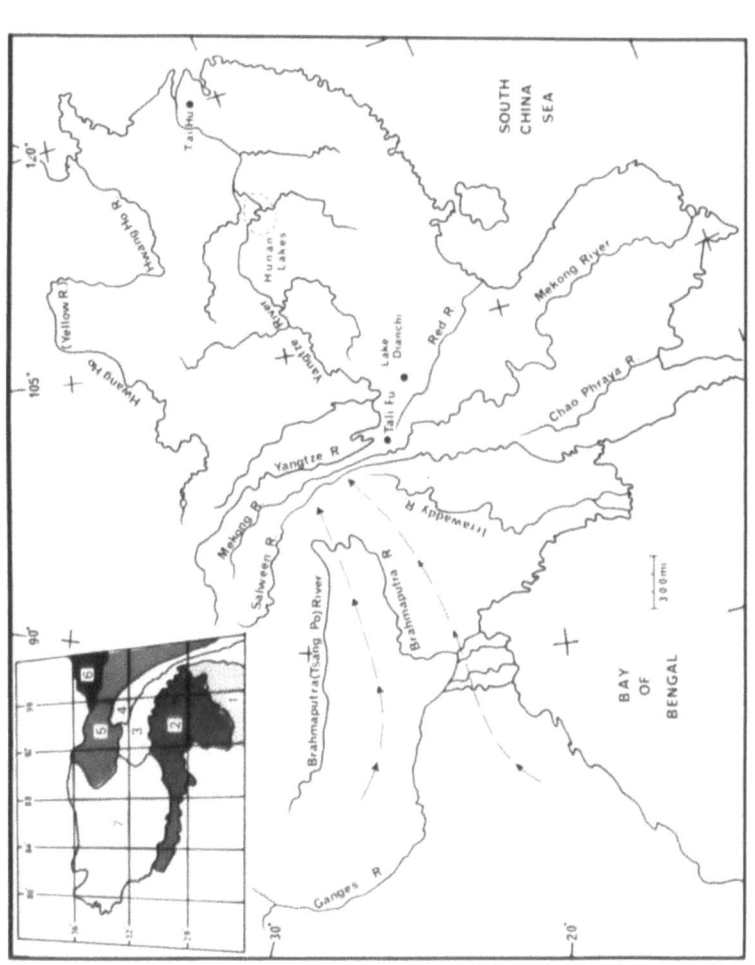

Figure 6.2. Major rivers of Asia. The inset (upper left) shows drainage patterns: (1) Irrawaddy River; (2) Brahmaputra River; (3) Salaween River; (4) Mekong River; (5) Yangtze River; (6) Hwang Ho (Yellow River); (7) Tibet Lake Region. The dashed lines with arrows define the probable route of introduction of faunal elements from the Indian Plate into Northern Burma and southwestern China. Tali Fu (= Lake Erhai = Er Hai = Dali Fu = Tali Fu of Dali county) drains into the Mekong River; Lake Dianchi at Kunming drains into the Yangtze River. The lake district in Hunan and Hubei (with *Lithoglyphopsis*) is circled. Adapted from Davis (5).

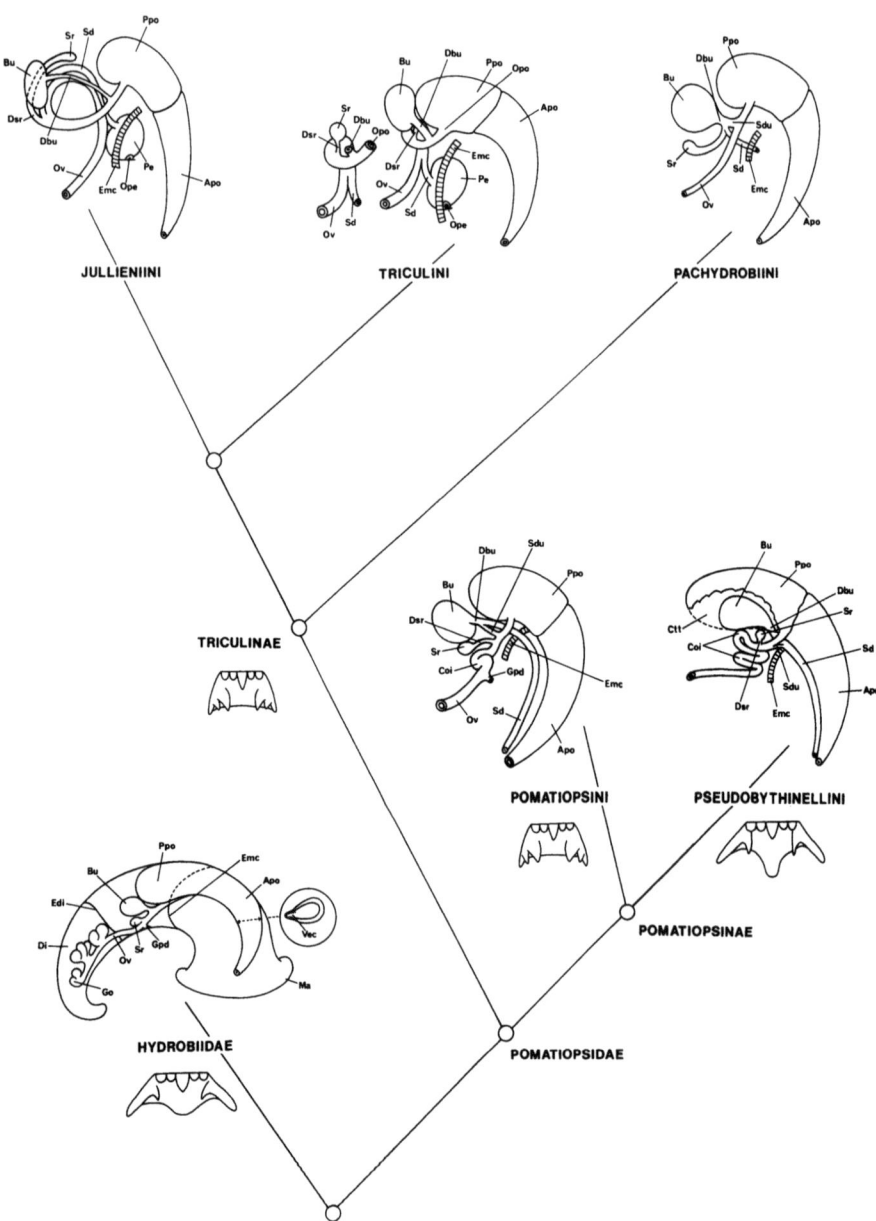

Figure 6.3. Cladogram showing the evolution of two families; the divergence of Triculinae and Pomatiopsinae; and the divergence of tribes. The female reproductive system anterior to the gonad and the central tooth of the radula are shown.

Apo = anterior pallial oviduct = capsule gland
Bu = bursa copulatrix
Coi = oviduct coil

(Legend continued on following page)

10. *Tricula bollingi* of Northwest Thailand and Yunnan, China (at the Mekong River) was considered (9) to have the most generalized (primitive) character-states of all Triculinae thus far studied. It was therefore more probable that the type species of *Tricula* (*Tricula montana* Benson, 1843) from northern India had the same generic-level morphology as *T. bollingi* than the morphology of the *T. aperta* group (Northwest Thailand to Cambodia; Mekong River drainage). It was also possible (1) that species of *Tricula* from India (genuine *Tricula*) had evolved derived character-states and were thus as divergent from the presumed primitive conditions as the *Tricula aperta* group, or (2) due to convergence Indian *Tricula* only resembled taxa from Thailand and China; they might actually belong to another family (Hydrobiidae?). If the last possibility was reality, the hypotheses on the origin of the Triculinae and the pomatiopsid-*Schistosoma* involvement were seriously weakened.

11. Given the impact of the Himalayan orogeny as a driving force of the spectacular triculine radiation, too few species of *Tricula* or species morphologically close to *Tricula* had been found. There were numerous species of *Pachydrobia, Lacunopsis, Hubendickia*, and others; only three species of *Tricula sensu* 1979 (5) had been found in Southeast Asia. It was predicted that more species of *Tricula* or related to *Tricula* would be found, some of which would transmit *Schistosoma*. Some of these species of *Schistosoma* would be new to science. The place to search was northern Burma and Yunnan, China.

◁─────────

Ctt = tissue of Ppo cut away to show the bursa
Dbu = duct of the bursa
Di = digestive gland
Dsr = duct of the seminal receptacle
Edi = anterior end of digestive gland
Emc = posterior end of the mantle cavity
Go = gonad
Gpd = gonopericardial duct
Ma = mantle collar
Ope = opening to pericardium
Opo = opening into the albumin gland
Ov = oviduct
Pe = pericardium
Ppo = posterior pallial oviduct = albumin gland
Sd = spermathecal duct
Sdu = sperm duct
Sr = seminal receptacle
Vec = ventral ciliated channel

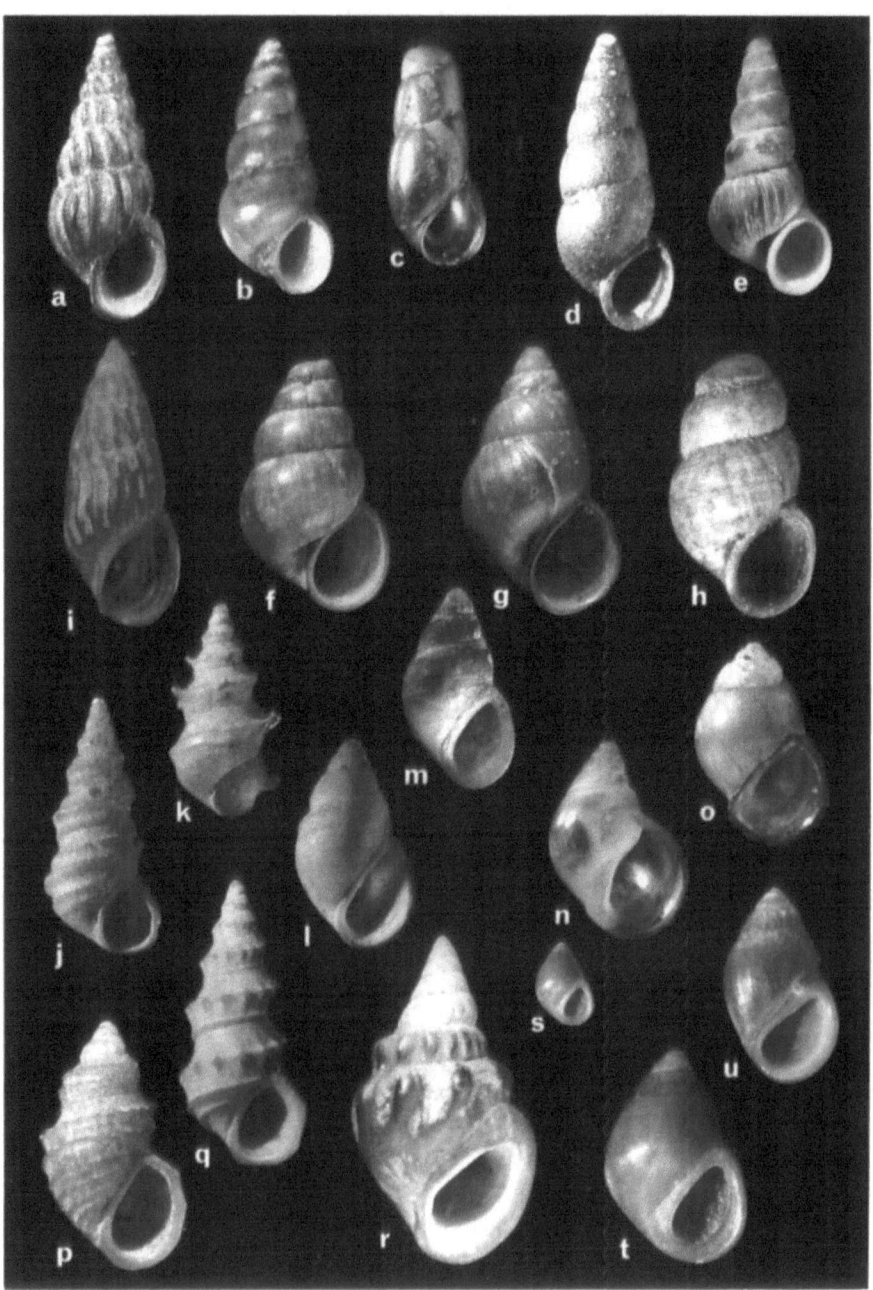

Figure 6.4. Shells of representative species of pomatiosine and triculine genera with their lengths given in millimeters. From Davis, (5).

a. *Oncomelania hupensis*, 9.2
b. *Pomatiopsis lapidaria*, 8.2

(Legend continued on following page)

In 1979 I listed (5, Table 16, p.113) 19 nominal genera that I considered to be Triculinae. Three of these listings are synonyms. I had anatomical data to objectively define only 11 of the 16 accepted genera. As of this review I have anatomical data for 20 genera of the 23 genera now considered to be Triculinae; seven genera were described as new since 1979. These additional data have permitted a more exacting analysis of cladistic relationships, the coevolution of schistosomes with taxa of some of these clades, biogeographic deployment, and ecology.

In this review I present an updated classification based on phylogenetic analysis, a discussion of character and character-states of importance when assessing snail phylogeny, the impact of snail ecology and phylogeny on the evolution of prosobranch-transmitted schistosomes, the origin of coevolved specificity, aspects of triculine ecology, and a biogeographic overview.

Classification

The outline of the current classification of relevant taxa is given below. The asterisk indicates that some species or subgeneric taxa of that genus transmit *Schistosoma*. I have not listed here the other seven genera of Pomatiopsinae variously located in South America, S. Africa, Australia, Japan, and the United States. For details of the genera see Davis. Shells of these diverse genera are illustrated in Figures 6.4–6.9. (5).

◁───

c. *Cecina manchurica*, 7.4
d. *Idiopyrgus brasiliensis*, 6.2
e. *Tomichia ventricosa*, 7.0
f. *Blanfordia bensoni*, 8.5
g. *Fukuia kurodai*, 8.8
h. *Coxiella striatula*, 5.5
i. *Hubendickia siamensis*, 6.2
j. *Karelainia davisi*, 4.4
k. *Karelainia davisi* (spiny phenotype)
l,m. *Hydrorissoia munensis*, 4.0; 3.5
n. *Neotricula aperta* (alpha race) 5.8
o. *Halewisia expansa*, 4.0
p. *Hydrorissoia elegans*, 5.0
q. *Neoprosthenia levayi*, 5.8
r. *Pachydrobia vaiabilis*, 13.5
s. *Pachydrobiella brevis*, 2.2
t. *Pachydrobia paradoxa*, 8.3

Family: Pomatiopsidae
  Subfamily: Pomatiopsinae
    Tribe: Pomatiopsini
      Genus: *Oncomelania** Gredler, 1881 (Fig. 6.4 a) (11).
    Tribe: Pseudobythinellini
      Genus: *Pseudobythinella* Liu and Zhang, 1979 (Fig. 6.6b,c) (12).
        "*Akiyoshia*" Kuroda and Habe, 1954 (13).
  Subfamily: Triculinae
    Tribe: Jullieniini
      Genus: *Hubendickia* Brandt, 1968 (Fig. 6.4i) (14).
        Hydrorissoia Bavay, 1895 (Fig. 6.4l,m,p) (15).
        Jullienia Crosse and Fischer, 1876 (Fig. 6.8i,j,l) (16).
        Karelainia Davis, 1979 (Fig. 6.4j,k) (5).
        Kunmingia Davis and Kuo, 1984 (Fig. 6.7a,b) (9).
        Neoprososthenia Davis and Kuo, 1984 (Fig. 6.4q) (9)
        Pachdrobiella Thiele, 1928 (Fig. 6.4s) (17).
        Paraprososthenia Annandale, 1919 (18).
        Parapyrgula Annandale and Prashad, 1919 (3).
        Saduniella Brandt, 1970 (Fig. 6.8a,c) (19).
    Tribe: Pachydrobiini
      Genus: *Gammatricula* Davis and Liu, 1990 (Fig. 6.7o,p) (10).
        Halewisia Davis, 1979 (Figs. 6.4o; 6.6a) (5).
        Jinhongia* Davis, 1990 (Fig. 6.7k,l) (20).
        Neotricula* Davis, 1986 (Fig. 6.4n; 6.6d) (21).
        Pachydrobia Crosse and Fischer, 1876 (Figs. 6.4r,t; 6.6h) (16).
        Robertsiella* Davis and Greer, 1980 (Fig. 6.5) (6).
        Wuconchona Kang, 1983 (Fig. 6.7m,n) (22).
    Tribe: Triculini
      Genus: *Delavaya** Heude, 1889 (Fig. 6.6f; 6.7c,d) (23).
        Fenouilia Heude, 1889 (Fig. 6.6g) (23).
        Lacunopsis Deshayes, 1876 (Figs. 6.8g; 6.9a,c,d) (24).
        Lithoglyphopsis Thiele, 1928 (Fig. 6.6e) (17).
        Tricula* Benson, 1843 (Fig. 6.7e–j) (25).
    Tribe: Incerte Sedis
      Genus: *Taihua* Annandale, 1924 (2).

Higher Taxa Defined

The family Pomatiopsidae (Gastropoda: Prosobranchia) was differentiated from the Hydrobiidae by Davis (5); two subfamilies of Pomatiopsidae were recognized, the Pomatiopsinae and Triculinae. The phy-

6. Evolution of Prosobranch Snails Transmitting Asian *Schistosoma* 155

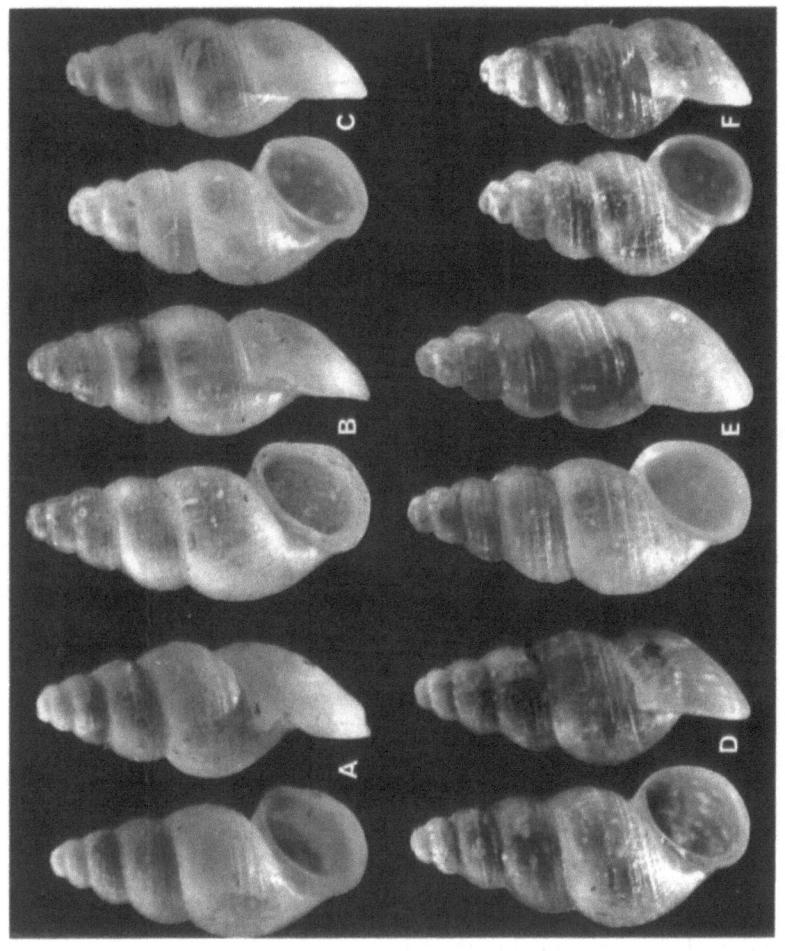

Figure 6.5. Shells of *Robertsiella kaporensis* (A–F) and *R. gismani* (D–F). Shell A is 3.12 mm long. Others are printed at the same scale.

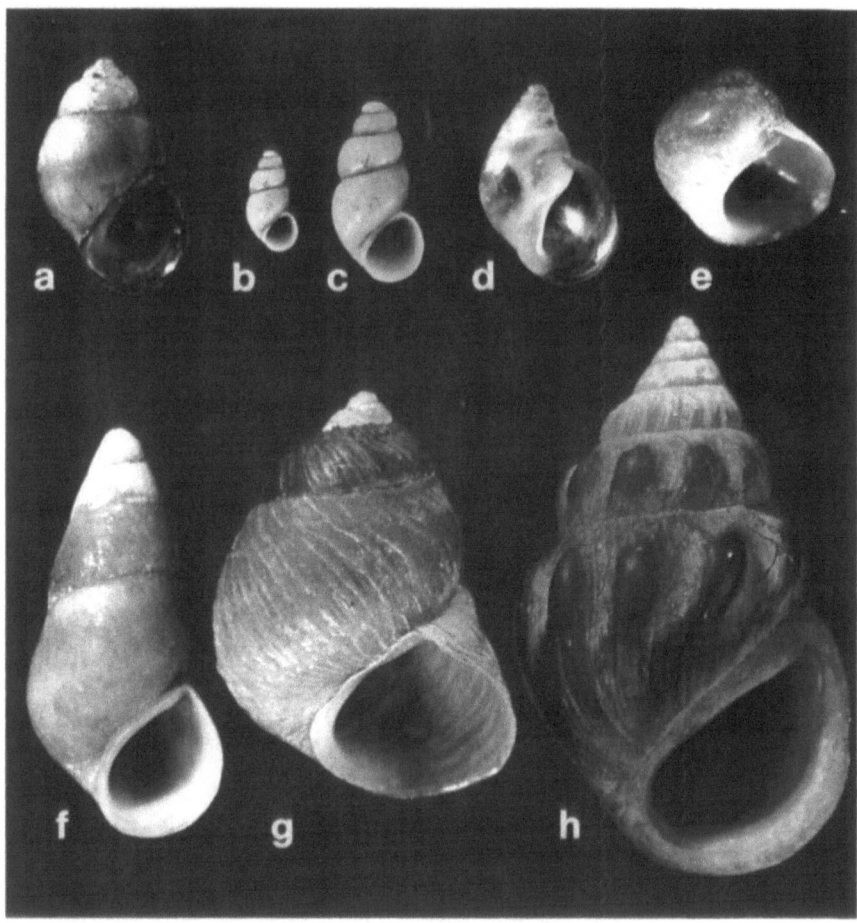

Figure 6.6. Shells of Pseudobythinellini and Triculinae. Shell a = 4.0 mm long; other shells are printed to the same scale.

a. *Halewisia expansa*
b,c. Species of *Pseudobythinella*
d. *Neotricula aperta*
e. *Lithoglyphopsis modesta*
f. *Delavaya dianchiensis*
g. *Fenouilia kreiteneri*
h. *Pachydrobia* sp.

logeny portraying the relationships among the higher taxa is given in Figure 6.3. The essential differences are found in the female reproductive system and central tooth of the radula. Note that in the Hydrobiidae the sperm travel into the female reproductive system via a cili-

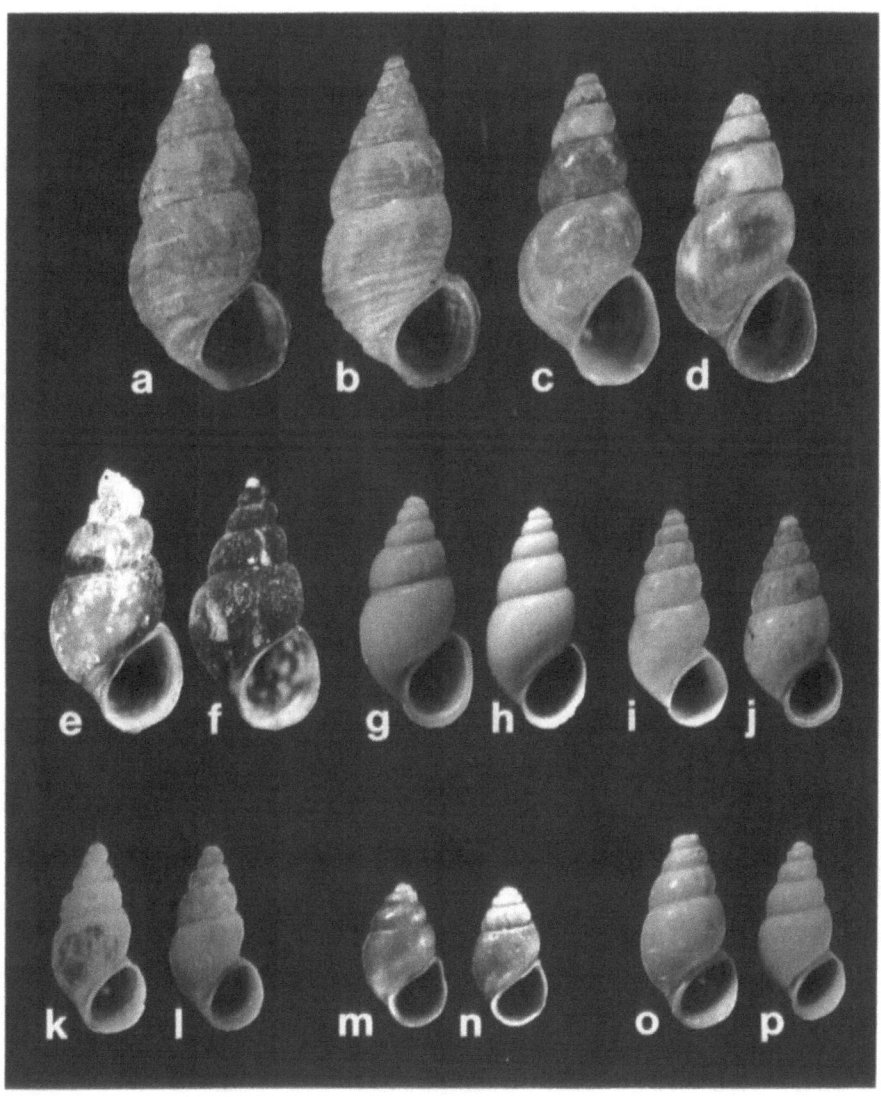

Figure 6.7. Shells of genera of Triculinae.

a,b. *Kungmingia kungmingensis*
c,d. *Delavaya dianchiensis*
e,f. *Tricula gregoriana*
g,h. *Tricula montana*
i,j. *Tricula bollingi*
k,l. *Jinhongia jinghongensis*
m,n. *Wuconchona niuzhuangensis*
o,p. *Gammatricula chinensis*

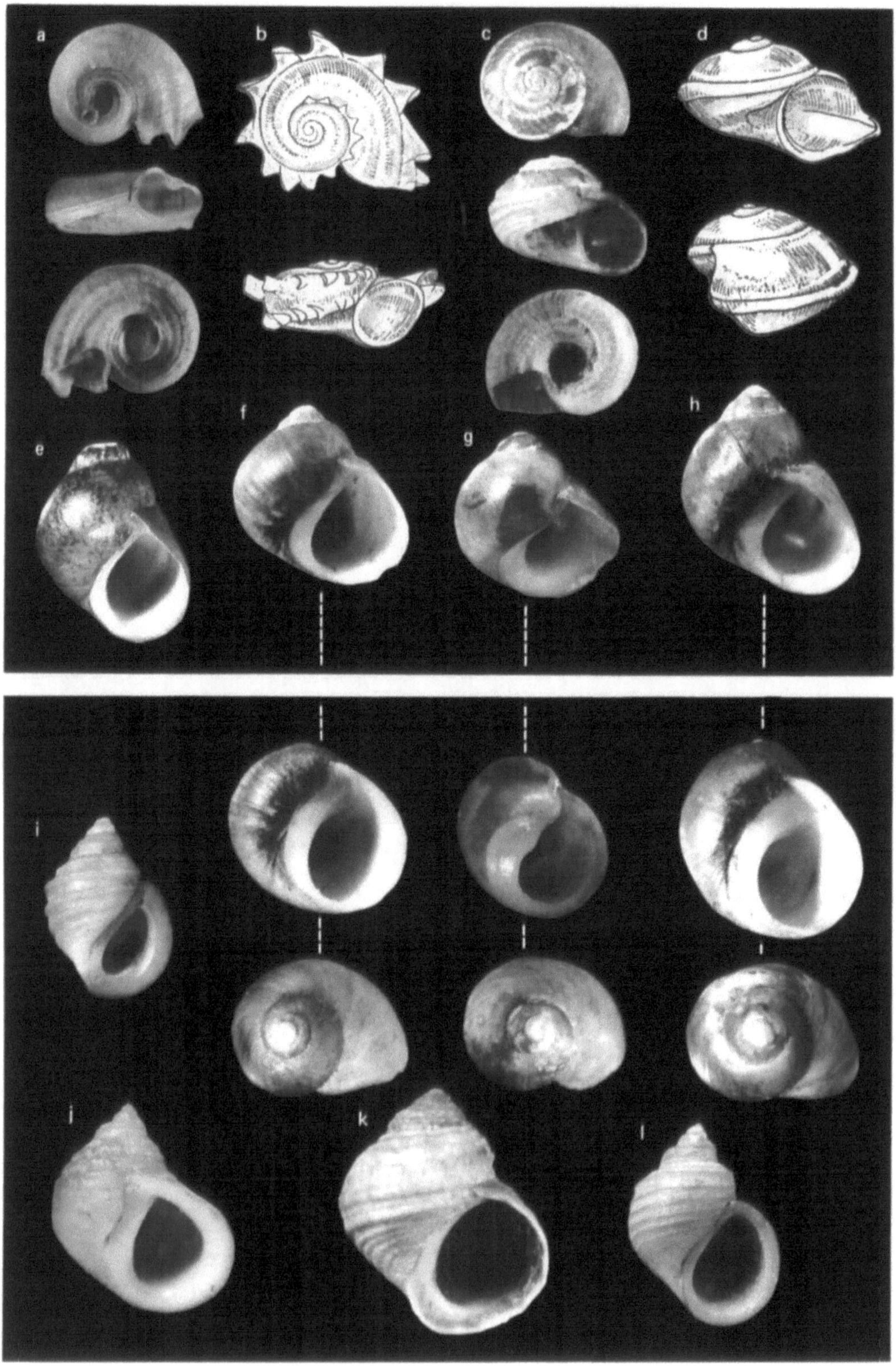

ated ventral channel (Vec) enclosed as a gutter in the pallial oviduct. In ontogeny this pallial oviduct starts as an organ open to the mantle cavity; it sutures up with further developement. In the Pomatiopsidae sperm pass through a closed spermathecal duct exterior to the pallial oviduct. During development the pallial oviduct opens a central passage by cavitation. It is important to understand that morphologically and developmentally (i.e., genetically) the Hydrobiidae diverged from the Pomatiopsidae prior to the Cretaceous, more than 130 million years ago. Some medical malacologists, e.g., Malek (26), still have not recognized the considerable differences between these families in time, space, or genetically. No Hydrobiidae transmits or can transmit a schistosome.

Both subfamilies of the Pomatiopsidae have a spermathecal duct, a pomatiopsid-type central tooth (square to rectangular with pronounced basal cusps arising from the face of the tooth (an exception includes the Pseudobythinellini discussed below), cover eggs (laid singly) with sand grains, and do not brood young. The penis is without complex glands or lobes. The differentiation of the pomatiopsid subfamilies from other higher taxa with a spermathecal duct was given by Davis et al. (27). These higher taxa are the Hydrobiidae: Littoridininae and the Amnicolinae.

The Pomatiopsinae are Gondwanian in distribution. Pomatiopsinae snails have an elongated spermathecal duct extending to the anterior end of the mantle cavity. In the tribe Pomatiopsini, the eyes are in pronounced bulges. There are a pedal crease, suprapedal fold, and an omniphoric groove. They progress by a step-wise mode. They have evolved from freshwater to an amphibious mode of existence, and in Japan *Blanfordia* has become terrestrial (5,7). *Cecina* has returned to the sea to live under cobbles at the high intertidal zone. The tribe

◁───

Figure 6.8. Triculine shells showing convergence with shells belonging to other families. Diameter is given in parentheses. From Davis (5), with permission.

a,c. *Saduniella planispira* (5.0 mm): Triculinae
b. *Pseudamnicola ornata* (1.5 mm): Hydrobiidae
d. *P. depressa* (1.5 mm): Lake Ohrid, Yugoslavia
e. *Anculosa subglobosa* (10 mm): Pleuroceridae, United States
f. *Lithoglyphus naticoides* (7 mm): Hydrobiidae, Europe
g. *Lacunopsis conica* (6 mm): Triculinae, Mekong River
h. *Lithoglyphus fuscus* (7.0 mm): Hydrobiidae, Europe
i. *Jullienia acuta* (6.3 mm): Triculinae
j. *J. crooki (7.6 mm): Triculinae*
k. *Littorina rudis* (9.2 mm): Littorinidae, United States
l. *J. flava* (7.3 mm): Triculinae

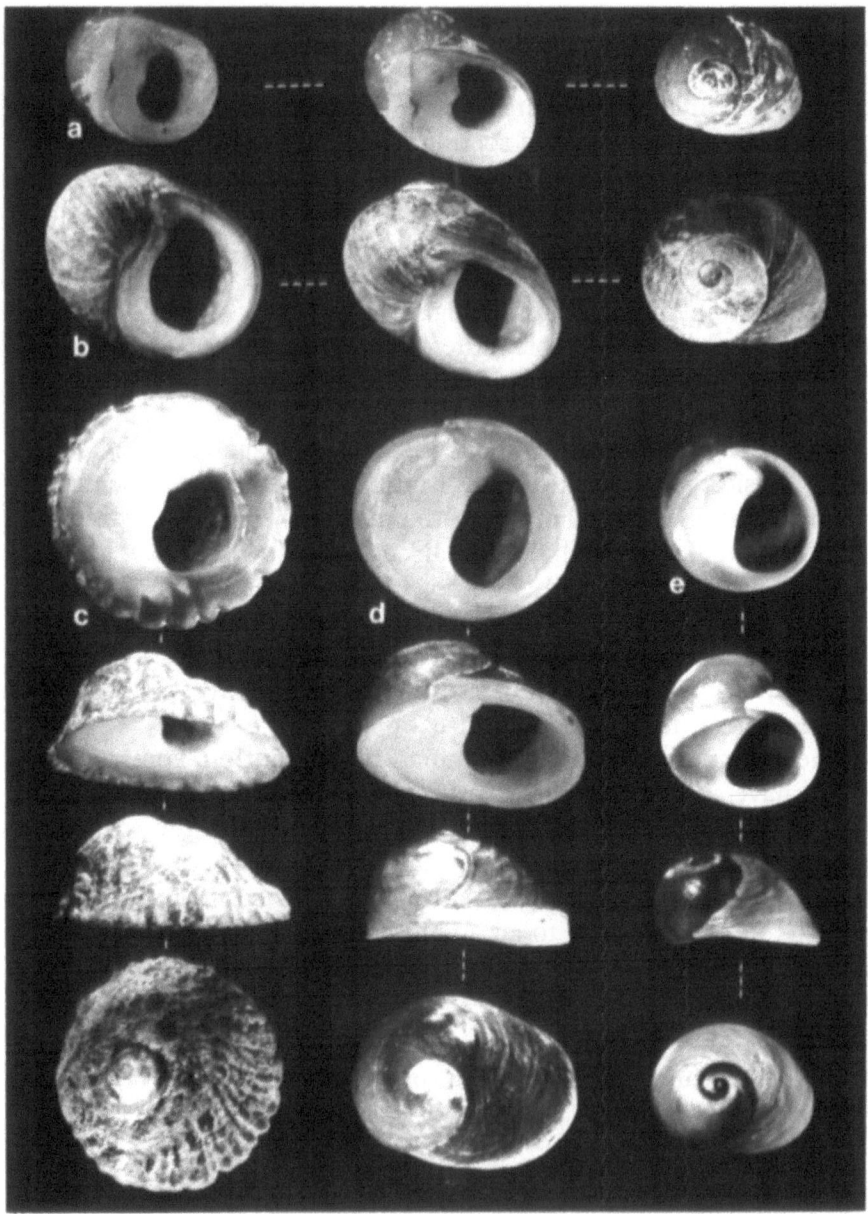

Figure 6.9. Shell convergence in three superfamilies. *Lacunopsis* (Rissoacea); *Littorina* (Littorinacea); *Spekia* (Ceritheacea). Diameter is shown in parentheses. From Davis (5), with permission.

a. *Lacunopsis globosa* (11 mm): Triculinae; Mekong River
b. *Littorina obtusata* (13.7 mm): Littorinidae, Marine, U.S.
c. *Lacunopsis fischerpietti* (16.3 mm): Triculinae, Mekong River
d. *Lacunopsis harmandi* (8.1 mm): Triculinae, Mekong River
e. *Spekia zonata* (11 mm): Pleuroceridae: Lake Tanganyika, Africa

Pseudobythinellini is placed in the Pomatiopsinae because of the elongated spermathecal duct. The central tooth is of the *Hydrobia* type, with a pair of cusps arising from the lateral angles. The eyes are not in pronounced eye lobes. There is no suprapedal fold, omniphoric groove, or pedal crease. Animals move by ciliary glide.

The Triculinae are Asian, with a distribution from northern India throughout South China and Southeast Asia. No Triculinae are found in Japan (Shikoku to Okinawa). Ioganzen and Starobogatov (28) referred their new genus *Sibirobythinella* to the Triculinae. However, given the extensive convergences in structure, such as the spermathecal duct and central tooth of the radula [reviewed by Davis et al. (27)], it is not at all certain that taxa from Siberia or northeastern Russia relegated to the Triculidae by Russian workers are Triculinae sensu Davis (1,5). *Sibirobythinella* is not, based on data provided, a triculine snail.

Triculinae snails have a short spermathecal duct opening to the posterior end of the mantle cavity. The snails are aquatic and do not have a pedal crease; most lack an omniphoric groove. In only a few derived genera are the eyes in pronounced lobes. Snails progress by ciliary glide.

Pomatiopsinae

Pomatiopsini

Type genus: *Pomatiopsis* Tryon, 1862 (29).
Type species: *Pomatiopsis lapidaria* (Say, 1817) (30) (see Fig. 6.4b; not type).
Designation: By monotypy.
Type locality: not recorded (Midwest, United States).
Ovate-conic to turreted shells. Shell length greater than 2.5 mm; most greater than 3.0 mm. Apical whorls are not flattened. Spermathecal duct runs from the bursa copulatrix to the anterior end of the mantle cavity. Spermathecal duct not tightly pressed to pallial oviduct. Sperm duct running posteriorly from oviduct to the bursa or the spermathecal duct at the bursa. The central tooth of the radula has prominent basal cups arising from the face of the tooth (Fig. 6.3). In China, there is only *Oncomelania*.
Type genus: *Oncomelania* Gredler 1881 (11).
Type species: *Oncomelania hupensis* Gredler, 1881, pp. 120–121, pl. 6, fig. 5 (11).
Type locality: U-tschang-fu, March 1879; = Hubei Province; Wu-Tshan-fu; Yen 1939 (31).
Designation: By monotypy.

Types: Bozen; lectotype and two paralectotypes. Bozen No. 89. No types at SMF (32).

Genera assigned (n = 8): *Blanfordia* (Japan) (Fig. 6.4f), *Cecina* (Japan/United States) (Fig. 6.4c), *Coxiella* (Australia) (Fig 6.4h), *Fukuia* (Japan) (Fig. 6.4g), *Idiopyrgus* (Brazil) (Fig. 6.4d), *Oncomelania* (China, Japan, Taiwan, Philippines, Sulawesi) (Fig. 6.4a), *Pomatiopsis* (United States) (Fig. 6.4b), *Tomichia* (South Africa) (Fig. 6.4e).

Oncomelania does not occur in Thailand, Burma, Laos, Cambodia, or Vietnam.

Pseudobythinellini

Type genus: Pseudobythinella Liu and Zhang, 1979 (12).
Type species: *Pseudobythinella jianouensis* Liu and Zhang, 1979 (12).
Designation: By original designation.
Type locality: Jian'ou, Fujian Province, P.R. China.
Type: Institute of Zoology, Academia Sinica, Beijing: No. FJ767701.
Synonymy: Erhaiini Davis et al., 1985, p. 69. (27).
Description. Ovate shells less than 2.5 mm long with flattened apical whorls. The sperm duct runs anteriorly from the oviduct to the spermathecal duct; it is tightly pressed to the pallial oviduct. Unlike the Pomatiopsini, the sperm duct enters the spermathecal duct far anterior to the bursa copulatrix; the duct of the bursa is thus elongated (the continuation of the spermathecal duct to the bursa from the point of entry of the sperm duct). Eyes are not in pronounced eye lobes. There is no suprapedal fold, omniphoric groove, or pedal crease. The central tooth is the *Hydrobia* or *Bythinella* type (Fig. 6.3); i.e., one or more pair of basal cusps arise from the lateral angle. Animals are aquatic and move by ciliary glide.

Discussion

The genus *Erhaia* was described (27) based on the data available for Chinese taxa described as *Bythinella*. *Pseudobythinella* Liu and Zhang, 1979 (12) was described as different from so-called *Bythinella* of China (1) because it had a tooth or node on the inner lip of the shell, and (2) the central tooth had two pairs of basal cusps, those on each side on the same level (not *Hydrobia*-like). As so-called *Bythinella* of China did not have the male reproductive system characters that serve to define *Bythinella* of Europe, a new genus, *Erhaia*, was described. However, it is now clear that in at least one species described as *Bythinella* from China [*B. chinensis* Liu and Zhang, 1979 (12)] a tooth on the columella can be observed when the aperture is tilted. Upon breaking open the shell it is seen that this "tooth" is the terminus of a thick, glassy, spiral columellar shelf or ledge. In summary, there appears to be no basis for placing in separate genera those taxa with an overall similar anatomical ground plan but where there is a tooth on the columella or not, or where the tooth is prominently seen in the aperture contrasted with the presence of a columellar thickening seen only upon breaking the shell. The type species of *Erhaia*, *E. daliensis* Davis and Kuo 1985 (27), has no columellar "tooth" in evidence in the aperture even when the aperture is rotated to examine as deeply as possible into the shell. However, upon breaking open the body whorl, a pronounced thickened columellar ridge is seen. This glassy ridge is also found in *Erhaia kunmingensis* Davis and Kuo, 1985 (27), but it is not very pronounced and forms no node.

The character of basal cusp position on the central tooth can be somewhat clarified. The two basal cusps on a side being on the same level or one above the other seems irrelevant to generic definition. Most species of *Pseudobythinella* (including Chinese taxa described as *Bythinella*) appear to have only one pair of basal cusps. They type species, *Pseudobythinella jianovensis* Liu and Zhang was described as having two basal cusps on each side, on the same level (side-by-side). "*Bythinella*" *chinensis* Liu and Zhang (12) was described with two basal cusps, the inner pair above the outer pair. Later, *Pseudobythinella shimenensis* Liu et al (33) was described having only a single pair of basal cusps (arising from the lateral angle).

Variation in cusp number and arrangement must be carefully documented. In the study of *Erhaia kunmingensis* most central teeth of most snails had a single pair of basal cusps. However, some central teeth had two pairs of basal cusps (27, fig 16F), the innermost arising from the face of the tooth and lower than the outer pair arising from the lateral angle. That this inner pair arise from the face of the tooth perhaps indicates the origin of this tooth type from the standard pomatiopsid tooth (contrast the hydrobiid type with basal cusps arising from the lateral angle). In *Erhaia daliensis* virtually all central teeth of all snails had a single pair of basal cusps arising from the lateral angle of the central tooth. However, on one tooth (at least) (27, fig. 9D) a tiny second pair, lower than the first, is seen arising from the lateral angles. Finally, examination of topotypical "*Bythinella*" *chinensis* revealed only one pair of basal cusps arising from the lateral angles.

Triculinae

Pachydrobiini Davis and Kang, 1990 (20)

Type genus: *Pachydrobia* Crosse and Fischer, 1876 (16).
Type species: *Pachydrobia paradoxa* Crosse and Fischer, 1876 (16).

Designation: By monotypy.
Type locality: Mekong River, Cambodia.
Diagnosis: Genera of Triculinae in which the spermathecal duct bypasses the pericardium and the oviduct does not make a closed 360-degree twist. With the exception of *Wuconchona*, there is a sperm duct. The plesiomorphic state is to have the seminal receptacle arise from the bursa or the duct of the bursa; the derived state is loss of seminal receptacle with the function of the seminal receptacle taken over by new structures or modifications of the structures.
Genera assigned (n = 7): Halewisia, Neotricula, Pachydrobia, Robertsiella, Jinhongia, Gammatricula, Wuconchona.

Triculini Davis, 1979 (5)

Type genus: Tricula Benson, 1843 (25).
Type species: Tricula montana Benson, 1843 (25).
Designation: By monotypy.
Type locality: Bhimtal, North India.
Types: British Museum of Natural History, London Nos. 1964426; 1964427.
Diagnosis: Genera of Triculinae in which the spermathecal duct enters the pericardium. The oviduct makes a 360-degree closed twist. In the plesiomorphic state the seminal receptacle arises from the oviduct; the derived state is loss of the seminal receptacle, with the function of the seminal receptacle taken over by derived structure attaching to the oviduct–spermathecal duct juncture. The seminal receptacle arises from the inside of the oviduct coil (contrast Jullianiini). The duct of the seminal receptacle is relatively short. There is no sperm duct.
Genera assigned (n = 5): Delavaya, Fenouilia, Lacunopsis, Lithopsis, Tricula.

Jullieniini Davis, 1979 (5)

Type genus: Jullienia Crosse and Fischer, 1876 (16).
Type species: Jullienia flava (Desheyes, 1876) (24).
Type locality: Koko and Prec Omphil, Mekong River, Cambodia.
Designation: By monotypy.
Genera assiganed (n = 10): Hubendickia, Hydrorissoia, Jullienia, Karelainia, Kunmingia, Neoprososthenia, Pachydrobiella, Paraprososthenia, Parapyrgula, Saduniella.
Diagnosis: Genera of Triculinae in which the spermathecal duct enters the pericardium. The oviduct makes an open 360-degree loop. The spermathecal duct arises from the outside surface of the oviduct

far posterior in the oviduct circle, and the seminal receptacle follows the contour of the oviduct circle. The seminal receptacle arises from the outer surface of the oviduct (contrast the Triculini and Pachydrobiini). There is a pronounced increase in length of the duct of the seminal receptacle in the evolution of taxa in this clade. The vas efferens becomes lost. The posterior vas deferens becomes increasingly elongated (Fig. 6.15).

Morphology and Clades

From the foregoing definitions it is clear that the higher taxa are defined on the basis of a diverse array of qualitative characters and character-states, most of which are illustrated here in Figures 6.3 and 6.15. Considering *Hydrobia* as a representative of the Hydrobiidae: Hydrobiinae and outgroup for the Pomatiopsinae, and the 20 genera of Triculinae for which anatomic data are available, there are 28 characters and 87 character-states (Table 6.1) that serve to establish the phylogeny in Figure 6.10. The data involve 18% shell characters; 46% characters of the female reproductive system; 25% characters of the male reproductive system; and 11% characters of the digestive system. The phylogeny involves 27 synapomorphies, 13 homoplasies, and one loss for a minimal tree length of 41 steps; it is noncomputer-mediated but based on a simple set theory solution of nesting taxa on the basis of unique and unreversed characters (34) and minimizing homoplasies. It is a new synthesis.

The details of the phylogeny of the genera of the Triculini and Pachydrobiini involving 11 genera is given in Figure 6.11 based on data provided in Table 6.2. A computer analysis was involved. Only *Lithoglyphopsis* is not treated, as the data set for this genus has not yet been published. However, those data (63) place *Lighoglyphopsis* in the Triculini, diverging from an ancestor to *Fenouilia*.

Power of Morphology

Why use comparative anatomic data? Why not use molecular genetics? Although I have a laboratory of molecular genetics, why have I not used these tools to distinguish among taxa? The answers are as follow.

1. When I began studies of species resembling *Tricula* on the basis of shell (35) there were no anatomic data for any species of genera now classified as Triculinae. Anatomic ground plans are the basis for mol-

Table 6.1 Character and character-states use to differentiate genera of the Triculinae from each other and members of the Hydrobiidae: Hydrobiinae; to enable a phylogenetic analysis.

Shell ($n = 5$, 18%; total $n = 28$)
1. Size: (a) small (2.5–3.0 mm); (b) large (to 8 mm)
2. Shape: (a) ovate-conic; (b) derived shapes (b_1 = planispiral; b_2 = globose; b_3 = trochoid)
3. Sculpture: (a) smooth; (b) ribs; (c) spiral rows of nodes, raised ridges; (d) spines
4. Symmetry: (a) symmetric; (b) asymmetric (flattened face, flattened base)
5. Thickness: (a) thin; (b) increasing thickness

Female reproductive system ($n = 13$; 46%)
6. Gonopericardial duct: (a) small and in some cases not functional (from oviduct to pericardium as in *Pomatiopsis, Hydrobia, Stenothyra*); (b) fully functional with adequate diameter and length to serve as a sperm conduit; (c) lost (as in the Pseudobythinellini)
7. Sperm conduit: (a) via ventral channel enclosed in the pallial oviduct (as in *Hydrobia*); (b) via spermathecal duct separated from the pallial oviduct (as in Pomatiopsidae)
8. Spermathecal duct: (a) elongated, reaching from the bursa copulatrix to the anterior end of the mantle cavity (Pomatiopsinae); (b) from pericardium to the oviduct or bursa = modified gonopericardial duct (Triculinae: Triculini, Jullieniini); (c) from posterior end of the mantle cavity to the duct of bursa thus bypassing the pericadium (Triculinae: Pachydrobiini)
9. Oviduct configuration at the bursa: (a) coiled spring-like dorsal to the bursa (e.g., *Hydrobia, Pomatiopsis, Oncomelania*); (b) makes a tight 360° twist (e.g., Triculinae: Triculini, Jullieniini—*Karelainia*); (c) makes a narrow-diameter 360° circle (e.g., *Kunmingia*); (d) makes a medium-diameter 360° circle (e.g., *Hubendickia*); (e) makes a wide-diameter 360° circle (e.g., *Jullienia, Saduniella Neoprososthenia, Hydrorissoia*); (f) makes an enormously wide 360° circle (e.g., *Pachydrobiella*); (g) makes irregular loops (Pomatiopsinae, Pseudobythinellini); (b_1) runs to the pallial oviduct without making a 360° twist (Triculinae: Pachdrobiini)
10. Seminal receptacle: (a) arises from inside surface of the oviduct anterior to the coil or twist (*Hydrobia*, Pomatiopsinae, Triculinae: Triculini); (b) arises from outside surface of the oviduct in the 360° circle (Triculinae: Jullieniini); (c) arises from the duct of the bursa or the bursa (Triculinae; Pachdrobiini); (d) lost, function replaced by new adaptations (some Triculinae: Triculini, Pachydrobiini)
11. Duct of seminal receptacle: (a) at 90° to the oviduct or angles away from the oviduct, duct of bursa (all *except* Triculinae: Jullieniini); (b) follows contour of oviduct circle/twist (Triculinae: Jullieniini); (c) lost (see 10d)
12. Spermathecal duct: (a) joins oviduct before the oviduct twist or at the beginning of the twist (Triculinae: Triculini); (b) becomes part of the 360° oviduct circle joining the oviduct nearly 180° around the loop (Triculinae: Jullieniini); (c) joins the duct of the bursa, or bursa (Triculinae: Pachydrobiini)
13. Length of duct of seminal receptacle: (a) short (*Hydrobia*, Pomatiopsinae, Triculinae: Triculini, Jullieniini—*Karelamia, Kummingia,* some *Hubendickia*); (b) elongated (Triculinae; Jullieniini—*Saduniella, Neoprososthenia, Hydrorissoia, Jullienia*); (c) elongated (Jullieniini—*Pachydrobiella*); (a_1) lost so seminal receptacle fuses to the oviduct; (a_2) seminal receptacle lost
14. Sperm duct: (a) with (Pomatiopsinae; Triculinae: Pachydrobiini); (b) without (Triculinae: Triculini, Jullieniini)
15. Seminal receptacle lost; function taken over by new structures: (a) secondary seminal receptacles arise outside the spermathecal duct at the oviduct-spermathecal duct junction (Triculini-*Lacunopsis*); (a_1) functions move to inside the duct systems; (b) encapsulated in the wrappings around the common sperm duct (*Robert-*

Table 6.1 *Continued.*

siella); (c) inside the swelling between the duct of the bursa and the spermathecal duct (*Jinhongia*); (d) inside the oviduct (*Wuconchona, Gammatricula*)
16. Duct of the bursa: (a) short (Jullieniini-*Kunmingia*); (b) nearly ductless (Jullieniini-*Karelainia*); (c) long (Jullieniini-*Hubendickia, Saduniella, Neoprososthenia, Hydrorissoia, Jullienia*); (d) very long (Jullieniini-*Pachydrobiella*)
17. Bursa copulatrix: (a) ventral to oviduct coil; (b) posterior to the oviduct circle complex; (c) surrounded by posterior arc of the oviduct circle complex
18. Gonad: (a) with lobes; (b) reduced lobes to form a sac when fully charged; (c) one or more finger-like tubes filling the digestive gland

Male reproductive system (n = 7; 25%)
19. Male gonad: (a) with series of lobes (*Hydrobia*, Pomatiopsinae, Triculinae: Triculini, Pachydrobiini, Jullieniini—*Karelainia*); (b) modified to large collecting ducts with or without terminal lobes; (b_1) with wide collecting ducts and finger-like terminal lobes (*Neoprososthenia*); (b_2) with wide collecting ducts and clusters of small terminal lobes (*Hydrorissoia, Jullienia*); (B_3) with wide collecting ducts and few wide terminal lobes (*Pachydrobiella*); (b_4) finger-like filling the digestive glands (*Hubendickia, Saduniella*)
20. Vas efferens: (a) has (*Hydrobia*, Pomatiopsinae, Triculinae: Triculini, Pachydrobiini, Jullieniini—*Karelainia*); (b) does not have (Jullieniini—all other genera)
21. Vas deferens (Vd_1 between gonad and seminal vesicle): (a) short (*Hydrobia*, Pomatiopsinae, Triculinae—Triculini, Pachydrobiini, Jullieniini—all *Karelainia*); (b) elongates to run to style sac; (b_1) runs to posterior end of style sac before turning posteriorly to coil as seminal vesicle; (b_2) runs to anterior end of style sac before turning posteriorly to coil as seminal vesicle
22. Seminal vesicle: (a) coils as a spring in a posteroanterior direction in the digestive gland dorsal to the gonad, posterior to the stomach; (b) coils as a mass or knot of tubes just posterior to the stomach or overlapping on the posterior chamber of the stomach; (c) coils as a small knot of tubes where the esophagus enters the stomach
23. Penis: (a) without a chitinous stylet; (b) with a chitinous stylet (Triculinae: Pachydrobiini—*Robertsiella*)
24. Ejaculatory duct: (a) none; (b) in base of penis; (c) in neck; (d) hypertrophy and massive in neck
25. Anterior vas deferens: (a) runs from prostate without coiling; (b) forms thick mass of coils in neck–head

Digestive system (n = 3; 11%)
26. Central tooth: (a) *Hydrobia*-like; (b) Pomatiopsid-like
27. Central tooth–anterior row of cusps: (a) three or more, usually 2(3)-1-(3)2; (b) *Saduniella* or *Neoprososthenia*-like = type II (5); (c) type III, i.e., large triangular cusp support and serrated dominant central blade (*Jullienia, Hydrorissoia*); (d) type IV, i.e., shallow cusp support and massive rounded, serrated central blade, no lateral cusps (*Pachydrobiniella*)
28. Stomach: (a) with caecal appendix (*Hydrobia*); (b) no caecal appendix (Pomatiopsidae)

Autapomorhies
Pomatiopsinae
29. Pedal crease
30. Step-like mode of progression

Pseudobythinellinae
31. Sperm duct: from oviduct to spermathecal duct anterior to or equal with position where oviduct enters the pallial oviduct (contrast all Pomatiopsini where sperm duct runs from oviduct to the bursa or spermathecal duct at or close to the bursa

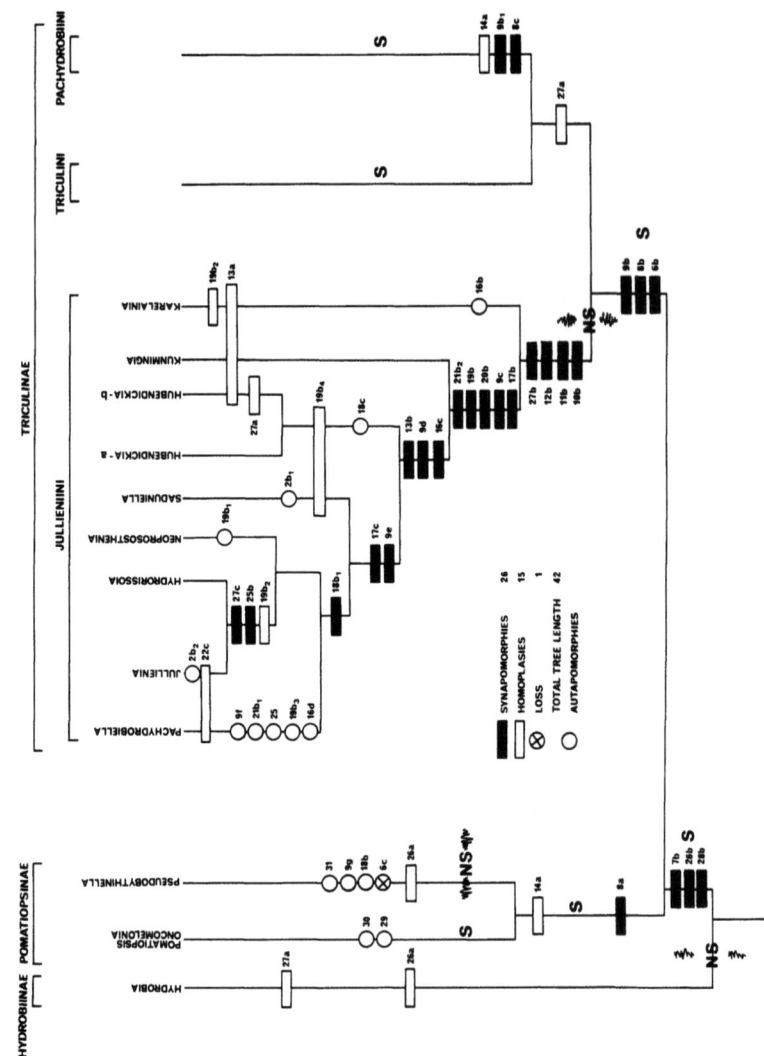

Figure 6.10. Cladogram based on a noncomputer analysis of data in Table 6.1. NS indicates clades not susceptible to infection with *Schistosoma* spp. Details of the Triculini and Pachydrobiini clades are given in Figure 6.11.

6. Evolution of Prosobranch Snails Transmitting Asian *Schistosoma* 169

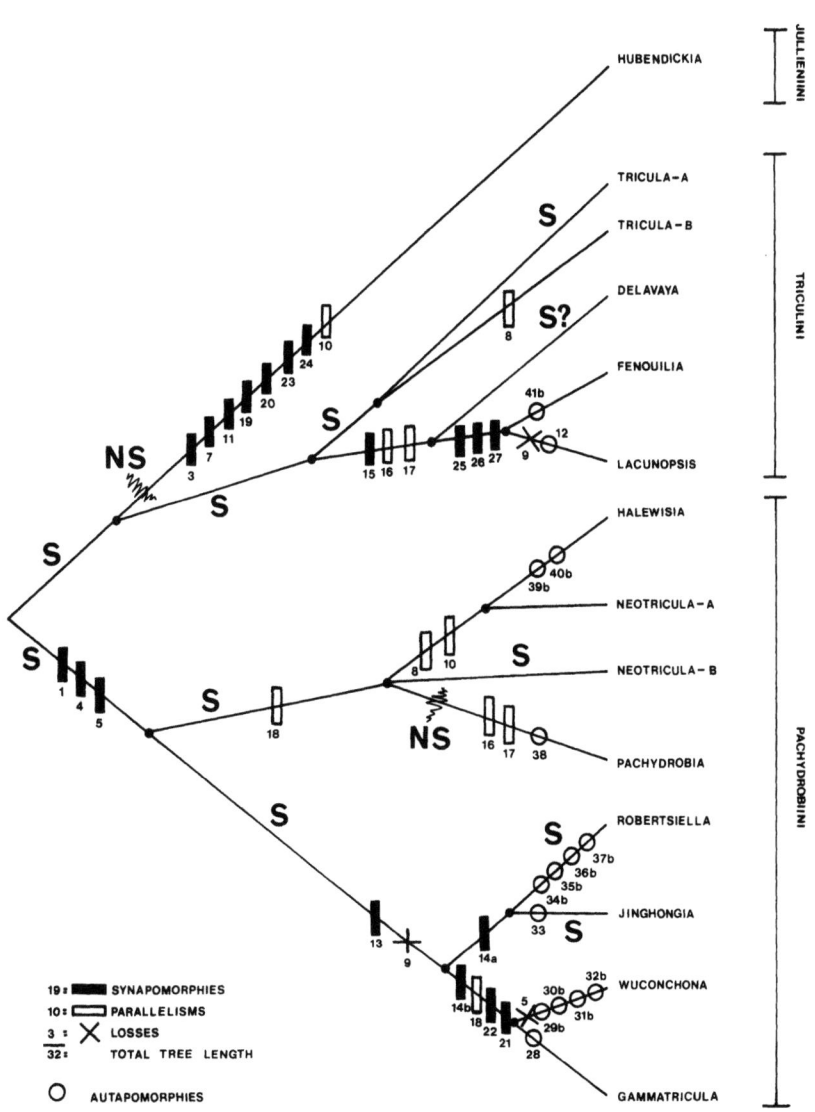

Figure 6.11. Cladogram. This phylogeny was constructed both with and without computer aid using the criterion of maximum parsimony. The computer program was Henning-86 version 1.5. (64) *Hubendickia* of the Jullieniini was the outgroup taxon. Characters were not weighted and character states were not given polarities. Multistate characters were broken down and scored in binary units. The "ie" option was used with no more than 100 trees to be retained in available tree space. The computer generated four equally parsimonious trees, one of which was congruent with the non-computer-aided cladogram. There were no differences among Hennig-86 maximum parsimony derived trees for the ordering of the outgroup or genera of the Triculini, nor was there disagreement about placing *Halewisia, Neotricula,* or *Pachydrobia* in one clade and *Robertsiella* and so on in another. Weighting one character, the position of the derived structure to replace the lost seminal receptable resolved the tree (10). NS = inability of species of the clade to be infected with or transmit *Schistosoma*. Adapted from Davis et al. (10).

Table 6.2 Characters and character-states used for the phylogenetic analysis.

Synapomorphies
1. Spermathecal duct opens to pericardium (0); to posterior mantle cavity (1).
2. Oviduct makes a closed, tight 360° twist posterior to the duct of the bursa (0,1).
3. Oviduct makes an open 360° circle (0,1).
4. Oviduct rune straight from gonad to the albumin gland or has a U-shaped bend (0,1).
5. Sperm duct is present (0,1).
6. Seminal receptacle arises from the right side of the oviduct posterior to the duct of the bursa (0,1).
7. Seminal receptacle arises from the left side of the oviduct posterior to the duct of the bursa (0,1).
8. Seminal receptacle arises from the duct of the bursa (0,1).
9. Usual seminal receptacle is lost, and its function is replaced by derived structures (0,1).
10. Duct of the usual seminal receptacle is long (0,1).
11. Duct of the usual seminal receptacle lies against the oviduct in an oviduct circle complex.
12. Derived seminal receptacle: two or more arise outside the juncture of the oviduct and the spemathecal duct (0,1).
13. Derived seminal receptacle arise inside the duct of the bursa, spermathecal duct joining the duct of the bursa, or inside the oviduct (0,1).[a]
14. Derived seminal receptacle arises (a) within the spermathecal duct/duct of bursa (0); (b) within the oviduct (1).[a]
15. Central tooth: anterior cusps five or more (0); one large triangular cusp (1).
16. Shell shape ovate-conic (0), derived (1).
17. Shell size small (0); derived is large (1).
18. Bursa elongated relative to length of albumin gland (0,1)
19. Male gonad is a simple tube (0,1).
20. Female gonad is one to three simple tubes (0,1).
21. Female gonad is a sac with a few tubes or, when filled with oocytes, a gorged sac (0,1).
22. Penis hase a strongly developed muscular white zone at concave edge (0,1).
23. Vas deferens leaves the anterior end of the gonad (0,1).
24. Vas deferens makes a U-turn at style sac to run posteriorly to form knotted seminal vesicle (0,1).
25. Head is squat (0,1).
26. Eyes are in pronounced lobes (0,1).
27. Foot is wide and powerful (0,1).

Autapomorphies
Gammatricula
28. Posterior half or third of the prostate is smooth, nonglandular (0,1).

Wuconchona
29. Shell columella (internal): (a) smooth (0); (b) raised spiral ridge (1).
30. Spermathecal duct: (a) long (0); (b) short (1).
31. Male gonad (a) with many lobes draining into vas efferens (0); (b) a wide sac, few lobes, floor of sac opens to vas deferens/seminal vesicle (1).
32. Dominant cusp of the lateral tooth: (a) approximately mid-tooth (0); (b) displaced toward the outside edge of the tooth (1).

Jinghongia
33. Derived seminal receptacle; a swelling in the juncture of the spermathecal duct— duct of the bursa (0,1).

Table 6.2. *Continued.*

Robertsiella
34. Derived seminal receptacle is a swelling in the duct of the bursa (0,1).
35. Penis: (a) without chitinous stylet (0); (b) with chitinous stylet (1).
36. Common sperm duct: (a) not encapsulated (0); (b) encapsulated (1).
37. Spermathecal duct: (a) without vaginal section (0); (b) with vaginal section (1).

Pachydrobia
38. Shell lip: (a) thin (0); (b) thick (1).

Halewisia
39. Spermathecal duct (a) joins duct of the bursa (0); (b) enters the bursa (1).
40. Sperm duct (a) short (0); (b) highly elongated (1).

Fenouilia
41. Head pigmentation: (a) melanin (0); (b) dusted with dense yellow granules (1).

Source: Davis et al. (10). With permission.
[a] There are corrections of two errors in Davis et al.'s 1990 Figure 14: 7b on the cladogram in the 1990 paper should be 13, as shown here with the adjusted line 13; 13a and 13b in Davis et al.'s 1990 paper should be 14a and 14b, as shown here with the adjusted line 14.

luscan classification, not genetic distance measures. It was, and still is, necessary to study the anatomical data for representatives classified in all nominal genera in India, Southest Asia, and southern Asia of what appeared to be rissoacean grade snails (and obviously not in the families Stenothyridae and Bithyniidae) to learn the actual number of genera that could be supported by the data and to assign species to genera and higher taxa. Genetic indices, by themselves, are inadequate for this task. For reasons continued below, the advisable course of action is to exhaust the anatomical data before investing in molecular genetics. This point is particularly true when, in the face of the cost of obtaining molecular data, the anatomical data (relatively inexpensively obtained) are highly varied in qualitative characters, when the data reveal groups of patterns, and when it is evident that additions of data from unstudied nominal genera and undescribed species will, with high probability, add significantly to the growing data base.

2. Early pursuit of anatomic data quickly revealed a richness of such data, showing both discrete patterns of ground-plan organization as well as clear-cut directions of evolution from ground plan a to a^1, $a^2 \ldots a^n$ Each species dissected added significant new data. It was clear during the 1970s that the return for a relatively small investment would be considerable by pursuing anatomic studies.

3. The Triculinae turned out to be an enormous monophyletic radiation. In 1979 I was primarily concerned with the immense adaptive radiation in the Mekong River, where a 300-mile stretch of river [Fig. 6. 12] contains what I then considered 11 genera and some 92 de-

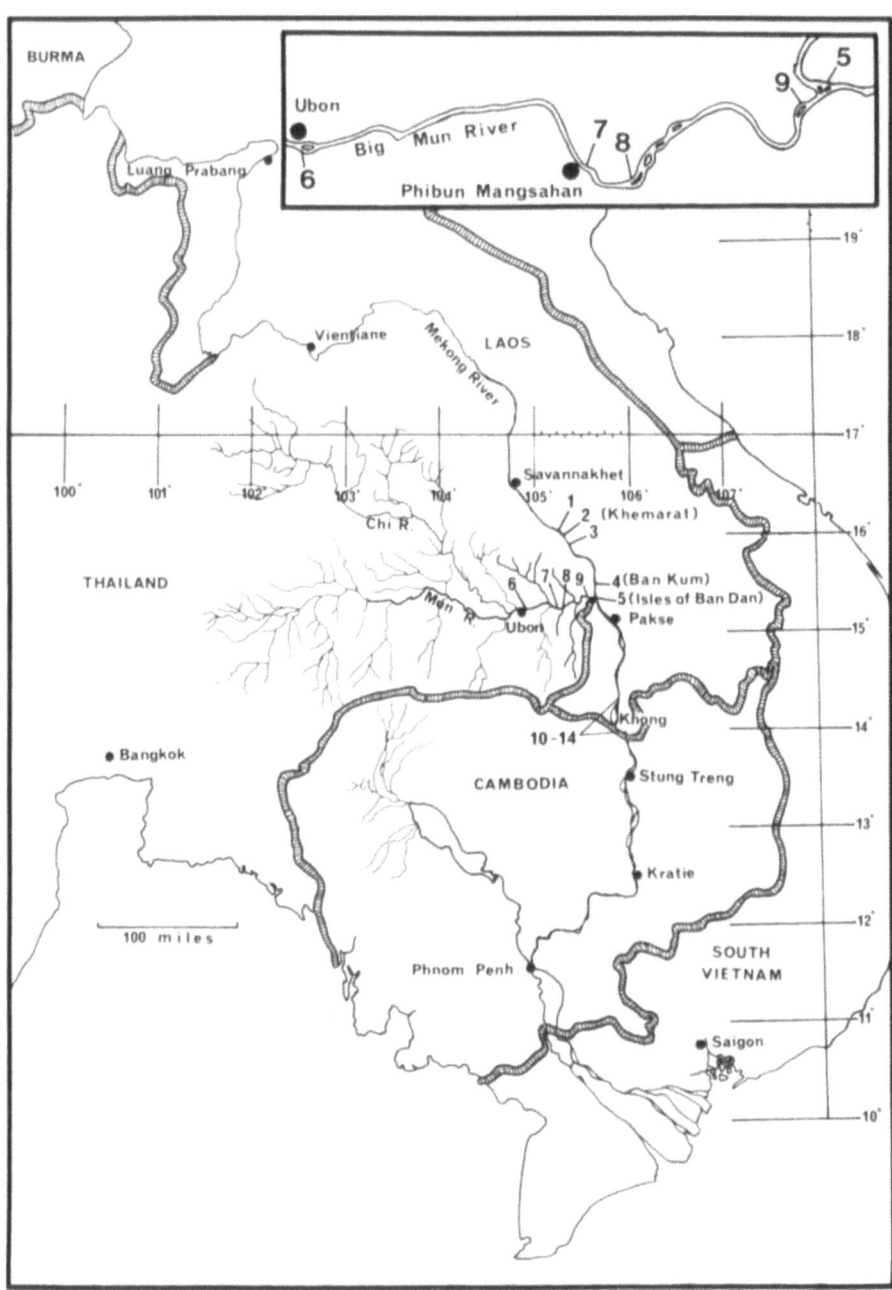

Figure 6.12. Localities rich in Triculine species diversity in the lower Mekong River and Mun River, studied by Davis from 1970 to 1975 (5). Triculines extend down to Kratie.

scribed species. More than a decade ago (1,33) the number of species exceeded 120 species (considering nominal taxa description in Burma, India, Vietnam, China, and other regions of Southeast Asia in addition to the Mekong River drainage. Research begun in Yunnan, China in 1983 and continuing today brings the number to more than 150 species. New genera are still being discovered and described. Considering the diverse shell pheotypes of what I consider undescribed species of Triculinae in the collections of the antiepidemic station in Kunming, Yunnan Province, and the rate with which new species are being found throughout the Yangtze River drainage, I estimate the number will exceed 200. The anatomic data from many of these species now being studied as well as what can be expected from many as yet unstudied species will add significantly to the database provided here.

4. During the 1970s and early 1980s (and continuing today) it was clear that the diversity of anatomic ground plans (mentioned above, point 2), coupled with the richness of qualitative anatomic character and character-states, would readily enable phylogenetic assessments. The cladograms that resulted with new discoveries were robust. Homoplasies have been readily detected.

Cladograms based on the anatomical data and the anatomical data are used in several ways. They are mapped on area cladograms of evolving river systems; and thus two hypotheses are examined for congruence. Do increasing degrees of anatomic specialization occur as one progresses down evolving river systems? A multivariate analysis of taxa versus characters and character-states yields a matrix of distance coefficients that may be considered relative genetic distances (1,5). The capacity of taxa to transmit *Schistosoma* may be mapped on area and anatomical-based cladograms to assess patterns of transmissions and coevolution. With these types of analyses in hand, the stage is set to make judicious choices of taxa and biogeographic locations to initiate molecular genetic studies. The time is now at hand to do so.

Vicariance, Dispersal, and Patterns of Speciation

An area cladogram is shown in Figure 6.13. The major stages discussed by Davis (5) are (1) India subtends the Asian mainland uplifting the Himalayan mountains from west to east. This uplifting initiates the drainages shown in Figures 6.2, 6.13 and 6.14. Ancestors of modern *Tricula*, *Oncomelania*, and *Schistosoma* are introduced to the Asian Plate through the Bhramaputra River gap. (2) There was an early period of lake formation and destruction as the new rivers began

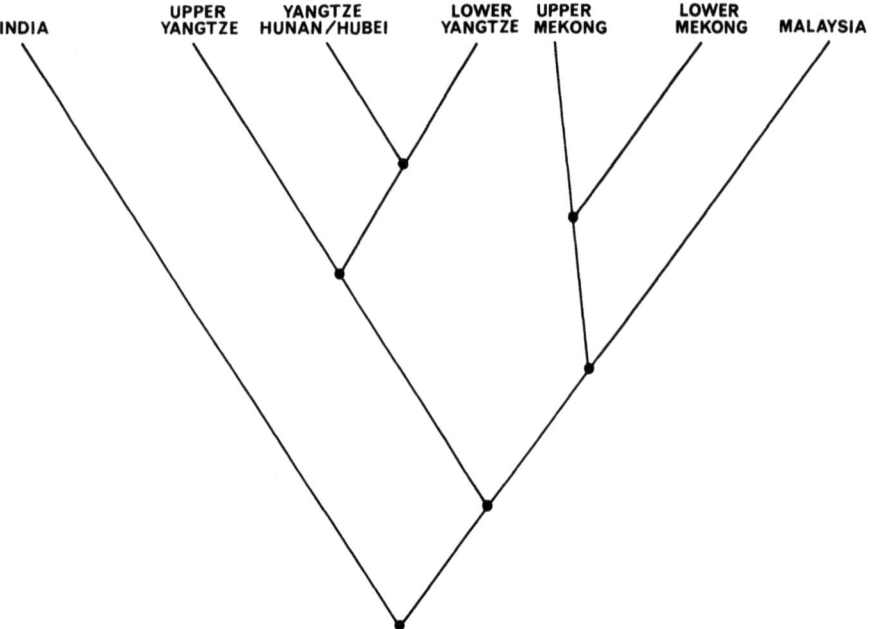

Figure 6.13. Area cladogram indicating the sequence of river evolution.

to evolve under the relentless Himalayan uplift some 12 ± 4 million years ago (18). Stream captures between the Mekong and Yangtze River occurred. (3) With lake lowering and complete isolation of the Yangtze and Mekong River drainages, new habitats suitable for snails evolved in the midriver regions. (4) Further evolution of the rivers continued, with shifts in suitable habitats moving to the lower rivers along with dispersal of faunal elements and the evolution of new taxa downstream. Modern distributions are the result of both vicariance and dispersal (5, 7, 36). The direction of dispersal and spurts of speciation have been down the evolving river systems.

Genera are listed in Table 6.3 under the appropriate tribe and grouped by a numerical ranking from 0 to 5, reflecting increasing grades of morphological specialization or synapomorphies (0 = most generalized in each tribe) These rankings are mapped on the river drainage systems (Fig. 6.14). A cladogram condensed from those in Figures 6.10 and 6.11 is also given. Several remarks are in order.

1. The Pomatiopsinae exist only in an arc from northern India through northern Burma and Yunnan, China throughout the Yangtze River drainage. *Oncomelania* does not now exist in India; it occurs in northern Burma as a fossil.

2. The three primary clades (tribes) of the Triculinae had evolved

6. Evolution of Prosobranch Snails Transmitting Asian *Schistosoma*

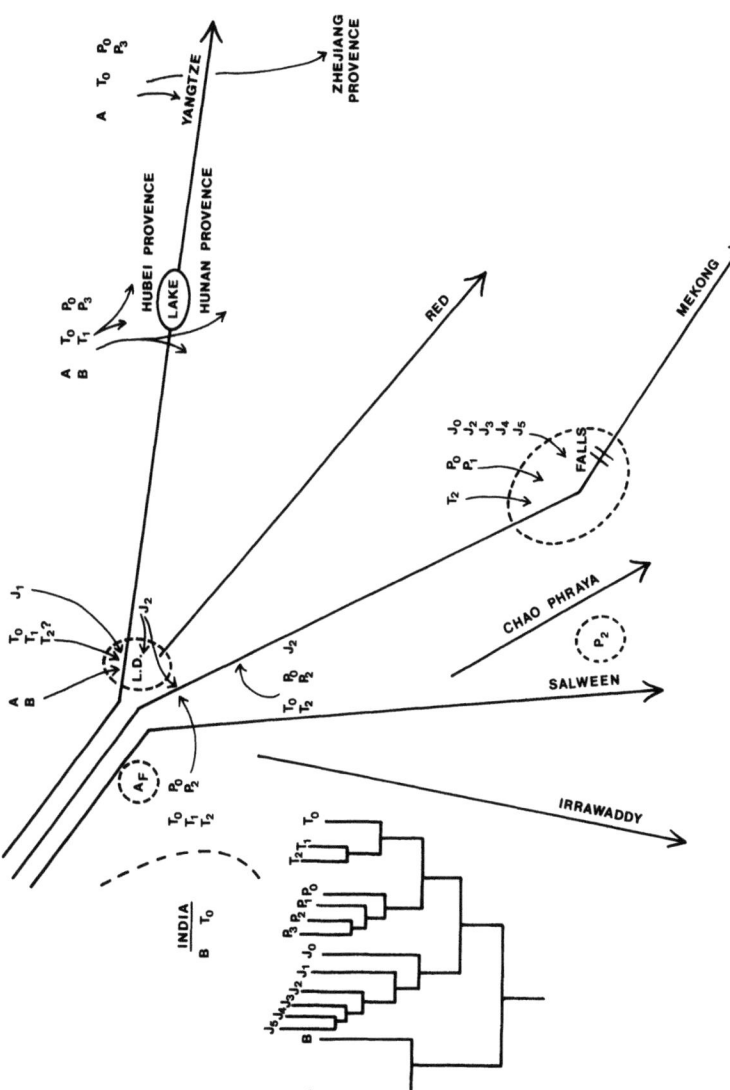

Figure 6.14. Cladogram showing generalized to derived genera (scale of 0 to 5) adapted from Figures 6.10 and 6.11 (see Table 6.3). Congruence of the area cladogram (Fig. 6.13), schematic relationships of the relevant rivers, and the distribution of faunal elements can be easily observed. Af = fossil *Oncomelania* in northern Burma; L.D. = lake district in Yunnan with lakes draining to the Mekong, Yangtze, Red, and Pearl Rivers. Rivers draining into the great Hunan river lake support *Lithoglyphopsis* (T_1). The area in the dashed circle around the falls of the Mekong River relates to the triculine rich regions numbered in Figures 6.12. Relate Figures 6.14 to Figures 6.2.

Table 6.3 Distribution of genera geographically.

Genus	India	N. Burma	Upper Yangtze	Mid Yangtze	Lower Yangtze	Upper Mekong	N.W. Thailand	Lower Mekong	Malaysia
Pomatiopsinae									
Oncomelania = A	–	F	+	+	+	–	–	–	–
Pseudobythinella = B	+	?	+	+	–	–	–	–	–
Triculinae									
Triculini = T									
Tricula 0	+	?	+	+	+	+	+	–	–
Delavaya 1	–	–	+	–	–	–	–	–	–
Fenouilia 1	–	–	+	–	–	–	–	–	–
Lithoglyphopsis 1	–	–	?	+	–	–	–	–	–
Lacunopsis 2	–	–	?	–	–	+	+	+	–
Pachydrobiini = P									
Neotricula 0	–	?	?	+	+	–	+	+	–
Halewisia 0	–	–	–	–	–	–	–	+	–
Pachydrobia 1	–	–	–	–	–	–	–	+	–
Robertsiella 2	–	–	–	–	–	–	–	–	+
Jinhongia 2	–	–	–	–	–	+	–	–	–
Wucochona 3	–	–	–	+	–	–	–	–	–
Gammatricula 3	–	–	–	+	+	–	–	–	–
Jullieniini = J									
Karelainia 0	–	–	–	–	–	–	–	+	–
Kunmingia 1	–	–	+	–	–	–	–	–	–
Hubendickia 2	–	–	–	–	–	+	+	+	–
Saduniella 3	–	–	–	–	–	–	–	+	–
Neoprososthenia 4	–	–	–	–	–	–	–	+	–
Hydrorissoia 4	–	–	–	–	–	–	–	+	–
Jullienia 4	–	–	–	–	–	–	–	+	–
Pachydrobiella 5	–	–	–	–	–	–	–	+	–

Numbers refer to scale of increasing anatomic specialization, with 0 = generalized, 5 = most derived. They are mapped over drainages in Figure 6.14; Letters are given in Figure 6.14, F = fossil.

before lake lowering and complete separation of the Yangtze, Mekong, Red, and Pearl [Nanpan River flows to Xi Jiang (=Si Kang) flowing to the Pearl] river systems had occurred. The presence of *Delvaya, Fenouilia*, and *Lacunopsis* in Yunnan, China indicates the complete spectrum of derived Triculini ground plans had evolved at an early stage (*Delavaya* and *Kunmingia* of the Yangtze system; *Fenouilia* in Lake Er Hai of the Mekong drainage. What we presume to be *Lacunopsis* is in both drainages).

3. *Neotricula* (P_o) of the Pachydrobiini is located in northwest Thailand, lower Mekong River, and mid to lower Yangtze Rivers. Studies to date have not yet shown *Neotricula* living in Yunnan. However, as mentioned above, given the large number of unstudied Triculinae known to exist in Yunnan, I fully expect species of *Neotricula* to be found there. Thus far, *Neotricula* seems to have diversified more in the mid to lower Mekong and Yangtze River drainages.

4. One might expect to find taxa of the most generalized type of each tribe (scored 0) to be found mid to lower river simply due to dispersal from northern Burma and Yunnan—and it has occurred. In the Jullieniini the most generalized taxon (*Karelainia*) is located in the lower Mekong River (Fig. 6.14). A taxon of this grade of organization has not yet been found in Yunnan, but, again, it is possible given the quantities of unstudied taxa there. However, both *Kunmingia* (ranked 1) and *Hubendickia* (ranked 2) do occur in Yunnan, China. The former is in the Yangtze River drainage; the latter has one species in the Pearl River drainage, another in the Mekong River.

5. Taxa ranked 3 or higher are found only at the mid to lower river localities; i.e., the most derived taxa are the farthest downstream. The Jullieniini dominate the lower Mekong River; the Triculini and Pachydrobiini dominate the mid to lower Yangtze River as well as Guangxi, Fujien, and Zhejiang Provinces (now isolated from the Yangtze River).

Indeed, increasing morphological specializations/innovations do occur as one progresses downstream. There is a different pattern of character and character-state change in each major clade with the most dramatic difference involving the female reproductive system, gonad of the male reproductive system, shell, and central tooth of the radula.

The Jullieniini are, given current data, absent from the Yangtze River drainage with the exception of *Kunmingia* living in a pool some 16 to 20 meters above Lake Dianchi that flows to the Yangtze River. Species of Triculini and Pachydrobiini are sprinkled throughout the Yangtze River drainage, with the most derived genera located from mid to lower Yangtze drainage. None comes from the Yangtze River itself, but they come from tiny mountain streams or from smaller rivers (e.g., *Lithoglyphopsis* in Hunan province). Both Pachydrobiini and the Triculini, the most derived taxa, have lost the usual seminal receptacle; the functions of the seminal receptacle have moved to unique locations, and the new storage facilities involve unique innovations. There is little differentiation in shell, radula, or male reproductive systems. For example, one cannot distinguish among *Tricula*, *Neotraicula*, or *Gammatricula* on the basis of shell (Fig. 6.7) or radula.

Only one genus of the Triculini has speciated extensively (>10 species) in the lower Mekong River, i.e., *Lacunopsis*, the most derived genus. Only three genera of the Pachydrobiini are found in the lower Mekong River (*Halewisia* with two species, *Neotricula* with one species, and *Pachdrobia*). Only *Pachydrobia* has speciated extensively (>10 species) in the lower Mekong. The lower Mekong River is dominated by the Jullieniini radiation (seven genera). Five of these genera have the most derived character-states of all Triculinae. The explosive Jullieniini radiation in the lower Mekong River involves progressive changes in the oviduct circle complex of organs of the femal reproductive system visually made clear in Figure 6.15. The trends involve an increasing diameter of the 360° oviduct loop, an increasing length of the duct of the bursa, an increasing length of the duct of the seminal receptacle, increasing length of the posterior vas deferens, modifications of the central tooth of the radula, increase in complexity of the male's anterior vas deferens, and increase in shell sculptural complexity and shell asymmetry.

These radiations occur in the Mekong River and major tributaries (e.g., Mun River) along a 300-mile stretch of river that overlaps the fall line (see dotted boundary in Figure 6.14). The river above this stretch has filled in with micaceous sand and thus no longer has suitable substrata for triculine snails. The river has cut deep into the Korat Plateau. The flat plateau has no permanent mountain streams that are suitable for snails of the tribes Triculini or Pachydrobiini as in south China. During the prolonged dry season the Korat Plateau is exceedingly dry.

Old Problems Resolved

A decade ago there were three areas of concern calling for resolution. (1) What was *Tricula* ss.? (2) Where were all the species of *Tricula sensu* 1980 that one might expect to find given the explosive triculine macroradiation (1)? (3) What were the anatomic character-state polarities involving the relation of the seminal receptacle to the oviduct

6. Evolution of Prosobranch Snails Transmitting Asian *Schistosoma* 179

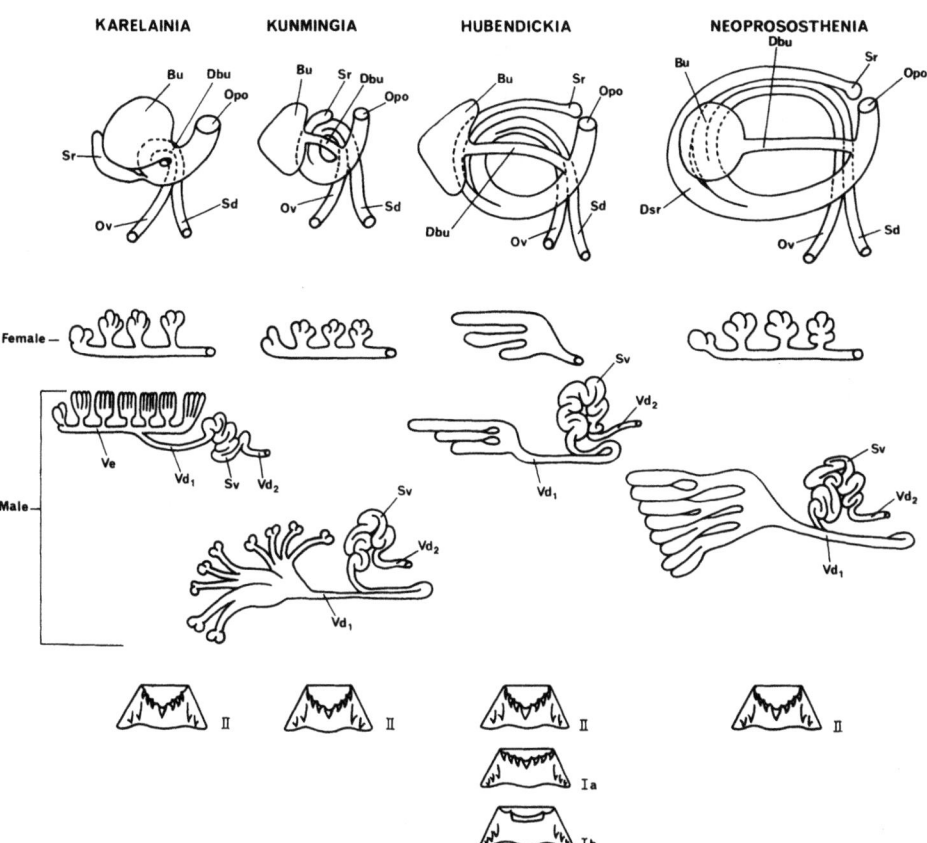

Figure 6.15. Illustrated character and character-states of the bursa copulatrix complex of organs, gonads, and radula for four genera of Jullieniini. The trends of opening up and expanding the diameter of the oviduct circle along with an increasing length of ducts of the bursa and seminal receptacle (from left to right) are obvious. The primitive-type male gonad with vas efferens (Ve) of *Karelainia* is shown. Only *Hubendickia* of the Jullieniini has species with a type I central tooth (type Ia is found in the generalized Triculini and Pachydrobiini, e.g., *Tricula* and *Neotricula*). Modified from Davis et al. (9).

Bu = bursa copulatrix
Dbu = duct of the bursa
Dsr = duct of the seminal receptacle
Opo = opening into albumin gland
Ov = oviduct
Sd = spermathecal duct
Sr = seminal receptacle
Sv = seminal vesicle
Vd_1 = posterior vas deferens
Vd_2 = anterior vas deferens
Ve = vas efferens

and to the bursa? What were the character-state polarities concerning the relation of the spermathecal duct to the bursa and to the end of the mantle cavity. It was essential to resolve these polarity questions to obtain a defensible phylogeny of the Pomatiopsidae.

1. *Tricula* question. For a decade it was clear that nominal *Tricula aperta* and *T. burchi* belonged to one genus and *Tricula bollingi* to another (1,6). The description of a new genus and assigning one or the other species groups to that genus depended on an analysis of the type species of *Tricula*, i.e., *T. montana* (25), a species from Bhim Tal of the lesser Himalayan mountains of northern India close to Nepal. Davis et al. (38) made an expedition to the type locality and adjacent regions in 1985 in order to resolve the problem.

The species was located; and the anatomy of *Tricula montana* was found to be identical to that of *T. gregoriana* of Yunnan, China (21) and similar to *T. bollingi* of northwest Thailand. Accordingly, *Lithoglyphopsis aperta* Temchaeron, later classified as *Tricula aperta* by Davis (5), the snail host of *Schistosoma mekongi* required a new genus. *Neotricula* was described for this species and *T. burchi* (37). Thus *Tricula* ss. exists in two arcs: one from northern India through northern Burma and Yunnan, China to the southeast coast of China are the other to northwestern Thailand.

2. *Tricula* diversity question. As discussed earlier, species resembling *Tricula* ss. with more generalized anatomies (*Tricula, Neotricula*) belong to the Tribes Triculini and Pachydrobiini. More than 90% of the estimated 60 relevant species occur in China sprinkled in the Provinces along the Yangtze River or Pearl River systems (Fig. 6.16). When species described as *Tricula* (on the basis of shell alone) from Burma and northern Vietnam are studied, it is expected that most of them will be either *Tricula* or *Neotricula*. The Triculini and Pachydrobiini morphostatic radiations considerably outnumber the Jullieniini adaptive radiations.

3. *Anatomic polarity question*. This question was partly resolved by Davis and Greer in 1980 (6), after the Davis 1980 (1) review was published. The question of the primitive connection of the seminal receptacle involved whether it was from the oviduct posterior to the duct of the bursa (as in *Tricula*) or from the duct of the bursa or common sperm duct (as in *Neotricula aperta*). The outgroup condition (seen in *Hydrobia* and *Oncomelania*) is the former. Davis and Greer (6), given the additional data from newly discovered *Robertsiella*, constructed two cladograms for the few relevant taxa for which data were available, based on each of the two assumptions. We then mapped the potential of each clade for transmitting *Schistosoma*. The most parsimonious solution indicated that the condition seen in the outgroup

6. Evolution of Prosobranch Snails Transmitting Asian *Schistosoma* 181

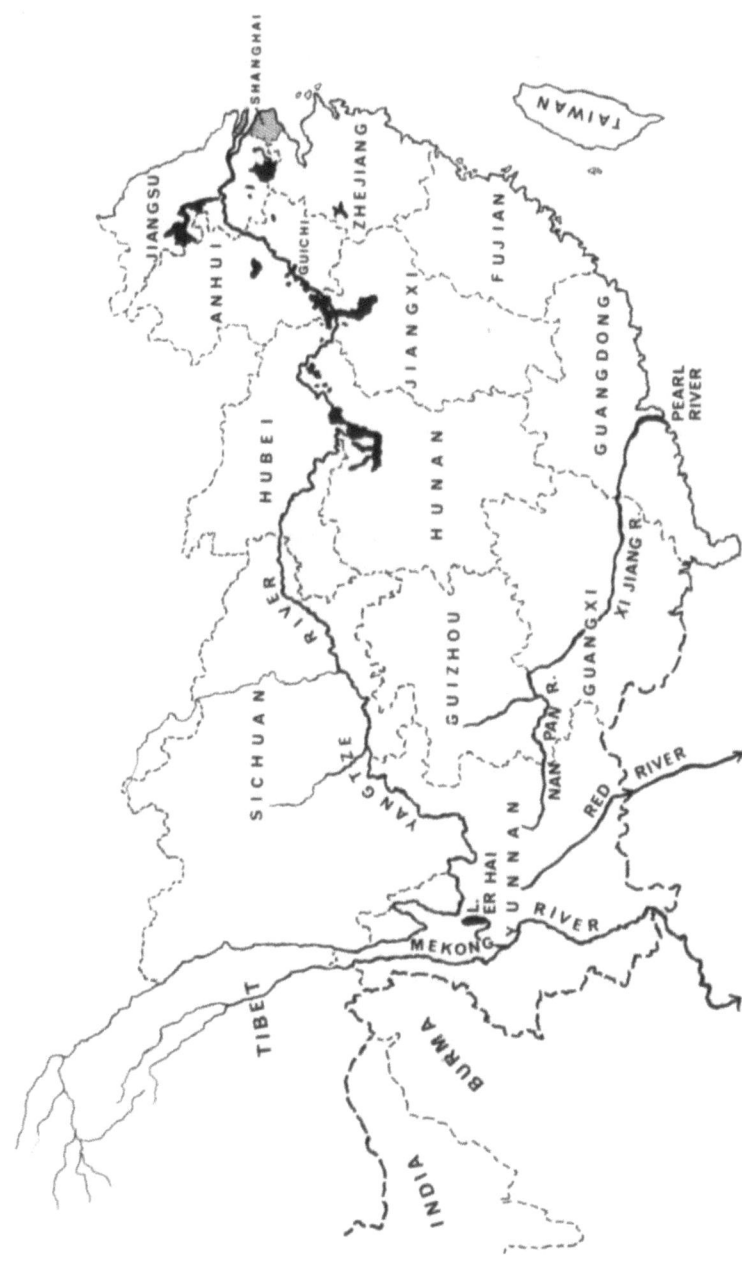

Figure 6.16. Provinces of southern China of considerable importance for studies of Triculini and Pachydrobiini.

was correct. This is strenghtened given the considerable data available today (Figs. 6.10 and 6.11).

The question of the origin of the spermathecal duct is more clearly ascertained today than it was a decade ago. The primitive condition is that seen in *Hydrobia* with a tiny gonopericardial duct connecting the oviduct to the pericardium at the region of the anterior style sac. It seems probable that the enlargement of the gonopericardial duct to form a spermathecal duct with entrance of sperm through an opening from the mantle cavity to the pericardium was the next derived step (37) followed by moving the spermathecal duct from the pericardium to open beside the pericardium at the rear of the mantle cavity, i.e., *Hydrobia* stage to *Tricula* stage to *Neotricula* stage.

Pomatiopsis retains the gonopericardial duct but also has a spermathecal duct runing from the bursa to the anterior end of the mantle cavity. It is thus probable that this spermathecal duct is not homologous with that of the Triculinae but developed by a different pathway (extension from the bursa). In both subfamilies the pallial oviduct opens a central passage by cavitation; there is no ventral ciliated groove as in *Hydrobia*. The two pomatiopsid clades were established before the disruption of Gondwanaland (note pomatiopsid genera in South America, South Africa, Australia, and China to Japan).

Coevolution

By coevolution I mean the term as defined by Janzen (39). The definition requires genetic specificity ad reciprocity; i.e., a character-state of a snail has evolved in response to a character-state of the parasite which character-state has itself evolved in response to an earlier character-state of the snail. That coevolution has occurred is convincingly documentable given all data available a decade ago, but the case is strengthened with data gathered over the past decade.

The facts are as follows [from the review of Davis (1)].

1. The *Schistosoma japonicum* complex (Table 6.4) involves worms of similar qualitative anatomy (40); the round eggs with minute spine are diagnostic. Parasites of this description are transmitted by a pomatiopsis genus (*Oncomelania*) and triculine genera (e.g., *Neotricula* and *Robertsiella*). Davis and Greer (6) recognized a *S. japonicum* complex in 1980 when, in describing the snail host for the schistosome infecting man in peninsular Malaysia (i.e. *Robertsiella kaporensis*), they argued that on the grounds of snail-schistosome coevolved specificity that the Malaysian schistosome (1) could not be either *S. japonicum* or *S. mekongi*, and (2) the Malaysian schistosome could be more closely related to *S. mekongi* than either was to *S.*

Table 6.4 Prosobranch snail hosts and the transmission of *Schistosoma* in Asia.

Schistosome	Snail host	Location
S. japonicum complex		
S. japonicum Katsuroda, 1904 (67)	*Oncomelania hupensis*	China, Japan, Phillipines, Sulawesi
S. mekongi Voge et al. 1978 (66)	*Neotricula aperta*	Lower Mekong River
S. malayensis Greer et al. 1988 (40)	*Robertsiella kaporensis*	Central Malaya
S. sinensium complex		
S. sinensium ss. Pao 1959 (48)	*Tricula* or *Neotricula* sp.	Sichuan, China
S. sinensium-like "a"	*Jinhongia jinhongensis*	Yunnan, China
S. sinensium-like "b"	*Tricula bollingi*	Northwest Thailand

japonicum. Yet, phenotypically the adult worms and eggs of the three species would closely resemble each other, i.e. they were cryptic species.

2. Given the *Oncomelania hupensis* paradigm discussed by Davis (1), we find today that one geographic strain of *S. japonicum* is not universally accepted by *Oncomelania hupensis* in other regions (41). For example, on Taiwan one population of *O. h. formosana* does not accept any strain of *S. japonicum*, another only a zoophilic strain, and yet another can experimentally transmit a human strain but does not do so in nature (1, 42). Allopatric subspecies of *Oncomelania hupensis* are readily hybridized in the laboratory. The results of mixing hybridization studies with infectivity studies show us that the potential to transmit a strain of *S. japonicum* is genetically controlled, few genes are involved, and the level of genetic specificity involving both snail and parasite genes operates today at the population level. The same pertains to *Schistosoma mansoni* and the pulmonate snail *Biomphalaria glabrata* (1).

That this level of specificity was derived from a state of lesser specificity is shown by the fact that *O. hupensis chiui* from an isolated corner of Taiwan can experimentally transmit all strains of *S. japonicum* (Chinese, Japanese, Taiwanese, Philippine, and probably Sulawesi) in the laboratory but is unexposed to any in nature. This population has been isolated from its progenetors and *S. japonicum* since Taiwan separated from the mainland about a million years ago. At some point prior to the divergence of the *Schistosoma japonicum* and *S. sinensium* lineages there must have evolved genetic-physiologic/chemical/cell-mediated immunity differences among higher taxa of snails and schistosomes such that *Schistosoma*-complex "X" could associate with pomatiopsid snails but not with Bithyniidae, Stenothyridae, Thiaridae, or pulmonates. *Schistosoma*-complex "Y" associated with some pulmonates (e.g., Bulininae) but could not use prosobranchs. At this stage all pomatiopsids probably were openly available for infection with *S.* complex "X" parasites; i.e., they were *Schistosoma*-naive within this association. It is from this fundamental platform of association that coevolved specificity arose.

3. Clearly this specificity evolved as the *Oncomelania hupenis* complex diversified and dispersed, and diverged from the Triculinae evolving macroradiation. The historical aspect of this divergence is tracked by simultaneously examining both drainage system cladograms and taxon cladograms (Figs. 6.10, 6.11, 6.13, 6.14). They are congruent. Genetically, the triculine-transmitted species of the *S. japonicum* species complex are isolated from the pomatiopsine-transmitted parasites of the *S. japonicum* species complex. As reviewed by Davis (1), attempts to attain reciprocal infections in the laboratory fail (43, 44). Once genetic isolation occurred, each *Schistosoma* lineage diverged along with the diversifying and speciating snail lineages.

Convergence, Phylogeny, and the Evolution of Schistosoma

As Cain and Harrison (45) pointed out, to establish a phylogeny one must work with a monophyletic assemblage. The monophyly of the Triculinae and Pomatiopsinae must be established if one is to use the phylogenies of these taxa to assess the patterns of coevolution with *Schistosoma*. To be sure of monophyly, convergences must be detected and convergent taxa removed from the analysis. If it can be done, then timing and direction of evolution must be determined making use of the fossil record, geologic events, and biogeographical analyses. If indeed these criteria can be accomplished, and if the anatomical database is robust, a credible phylogenetic hypothesis may be forged (46). Are Asian Triculinae monophyletic, evolving in space devoid of the Hydrobiidae?

Davis (5) discussed the major problem of convergence. Asian Triculinae were previously classified as Hydrobiidae primarily because *Lithoglyphopsis* from Hunan, China resembled *Lithoglyphus* from Europe on the basis of shell (Figs. 6.8, 6.9) and radula. This problem has finally been resolved; *Lithoglypus* has the anatomy indicating membership in the Triculinae: Triculini clade. On the basis of detailed anatomical data for numerous taxa, there is no evidence for any member of the Hydrobiidae in India, Southeast Asia, or southern China.

As Cain (47) pointed out, the phenotypic expressions we see are primarily the result of adaptations of organisms to their environment. Shells are profoundly affected in this way. For example, two snail species of different phylogenies may have nearly identical shells because they live in rocks in rapidly flowing water. This type of convergence is seen in Figures 6.6, 6.8, and 6.9, where similar looking shells belong to snails of three considerably divergent superfamilies. This type of situation is involved in the *Lithoglyphus–Lithoglyphopsis* convergence.

Minor problems of convergence and parallelism are not difficult to assess if there are sufficient taxa and the anatomical database is rich in characters and character-states, conditions obtained by the Triculinae. A case of convergence in the Triculinae is the loss of the seminal receptacle by one or more taxa in each of two clades; the Triculini and Pachydrobiini (Fig. 6.11). The overall data support the division of these tribes and the derived status of the genera of concern. Furthermore, data support the assertion that loss and function replacement arose independently in both tribes. In *Lacunopsis* of the Triculini, the function of sperm storage is taken over by several accessary receptacles branching to the exterior of, but communicating with, the spermathecal duct-oviduct juncture. In the derived genera *Robertsiella*, *Wuconchona*, and *Gammatricula* of the Pachydrobiini, the function is moved into the inside of the spermathecal duct and subsequently shifts to the inside of the oviduct (10, 20).

A case of parallelism (structure or character-state inherited from a common ancestor) is seen in the length of the duct of the seminal receptacle. In the primitive condition it is short. In the Jullieniini the short condition is maintained in *Karelainia* and is inherited by basal *Hubendickia*. Derived *Hubendickia* species have a considerably elongated duct, a condition obtained in all other (and more derived) genera of the Jullieniini. There is also convergence in this character. *Halewisia* of the Pachydrobiini has a considerably elongated duct.

With these problems of convergence resolved, the phylogeny of the Triculinae (Figs. 6.10; 6.11) meets the criteria of Cain and Harrison (45). It is robust. One can then map on this cladogram the actual capability of a taxon to transmit *Schistosoma*, such as Davis and Greer (6) did previously, and trace back to the roots the pathways of coevolution (Fig. 6.10). When one does this, it becomes clear that the snail-schistosome interaction is now phylogenetically constrained, and has been for a few million years.

Schistosoma sinensium *Species Complex*

It has become evident that two species complexes of *Schistosoma* transmitted by prosobranch snails have evolved in Asia. The *S. japonicum* complex infects snails of both the pomatiopsine and triculine clades. *Schistosoma sinensium* complex species infect, as far as is known, only triculine snails.

Schistosoma sinensium Pao, 1957 (48) is little known and poorly understood; it was described from Sichuan Province, China from a triculine snail. The snail species is neither *T. humida* nor *T. gregoriana* as reported (49, 50); the species of either *Tricula* or *Neotricula* is unknown. I found a mammalian-type *Schistosoma* in the type population of *Tricula bollingi* from Fang District, northwestern Thailand, that I classified as *Schistosoma* sp, (35). This population was subsequently studied by Kruatrachne et al. (51), Baidikul et al. (49), and Greer et al. (52). The odd-shaped egg with a sharp lateral spine much more resembles the egg of *S. mansoni* than that of *S. japonicum*. In a careful study of the morphology of the adults, eggs, and cerceria, Greer et al. (52) concluded that on the basis of the eight informative character-states studied, the *S. sinensium*-like species more closely resembled *S. japonicum* (62%) than *S. mansoni* (38%). The few allozymic data available (53) comparing the Thailand population of the *S. sinensium*-like parasite with *S. japonicum* show no similarity.

Jinhongia jinghongensis from the upper Mekong River drainage in Yunnan, China is reported to transmit a *S. sinensium*-like species (Guo, personal communication). Thus the few data available show a pattern of triculine transmission of *S.sinensium* and *S. sinensium*-like species in the hilly streamlets of the upper Mekong and Yangtze River drainages (Fig. 6.17; see also Fig. 6.21). Mammalian *Schistosoma* of undetermined species-group affinity has been found in *Tricula* and *Delavaya* of the Yangtze River drainage (21) and from Hunnan, China (Davis et al.) (63). They may belong to the *S. sinensium* complex.

Clearly much work needs be done to examine hypotheses here stated based on the scant data available: (1) There is a *S. sinensium* complex; the type population of *S. sinensium* from Sichuan Province is probably a different species than the taxon from northwestern Thailand. It is doubtful that the *S.sinensium*-like species transmitted by *Tricula* (Thailand) is the same species as the *S.sinensium*-like species transmitted by a different genus, i.e., *Jinhongia*. Remember the long-standing arguments pre-1970 about the Mekong schistosome being (or not being) *S. japonicum*. Worms and eggs were similar at first glance; however, careful study revealed differences in allometry

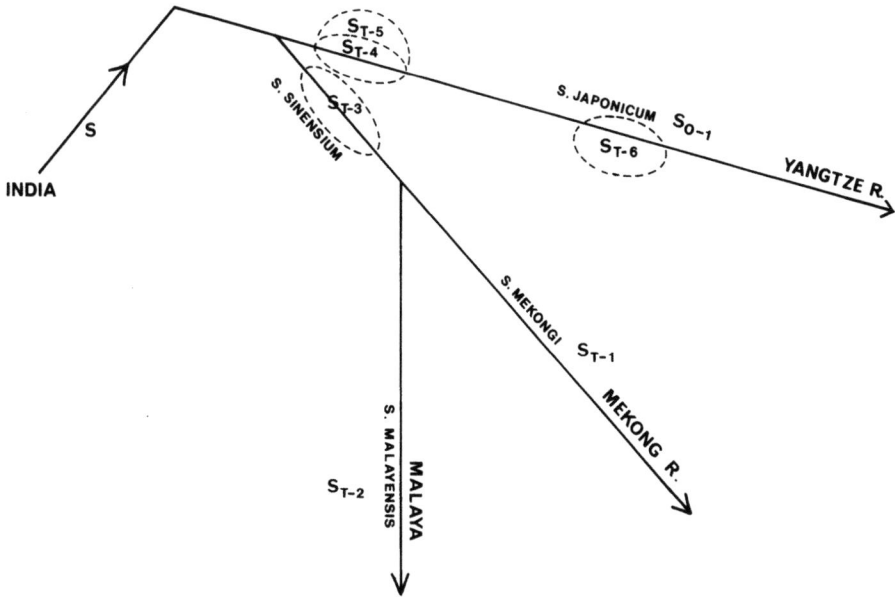

Figure 6.17. Area cladogram imposed on river systems with relative position of triculine taxa transmitting *Schistosoma* spp. S_{T-1} to S_{T-6} indicates *Schistosoma* species 1 to 6 transmitted by triculine snails. S_{T-3} to S_{T-6} are all possibly *S. sinensium*-complex species or populations. S_o = *Schistosoma japonicum* transmitted by *Oncomelania* S_{o-1} = one of several possible species in the complex transmitted by *Oncomelania*. As the identity of the triculine transmitting S_{T-6} has not yet been established, S_{T-6} is not treated in Figures 6.18 and 6.19.

among others. Molecular genetic studies proved conclusive (54). Just because the eggs from different regions and drainages look similar does not mean the taxa involved are the same species (40). (2) Taxa marked S_{T3-6} in Figure 6.17 are probably taxa referrable to the *S. sinensium* complex. (3) The *S. sinensium* ancestor was probably introduced into the headwaters of the Mekong-Yangtze Rivers already divergent from the *S. japonicum* complex ancestor (as indeed were *Indoplanorbis* and terminal spined schistosomes, i.e., the pulmonate snail-*Schistosoma* pattern of transmission found in India and Southeast Asia).

Coevolved Specificity Involving Triculine Snails

A model for coevolved specificity is given in Figures 6.18 through 6.21. Our knowledge for *S. japonicum* trasmitted by *Oncomelania hupensis* complex snails is much better understood owing to numer-

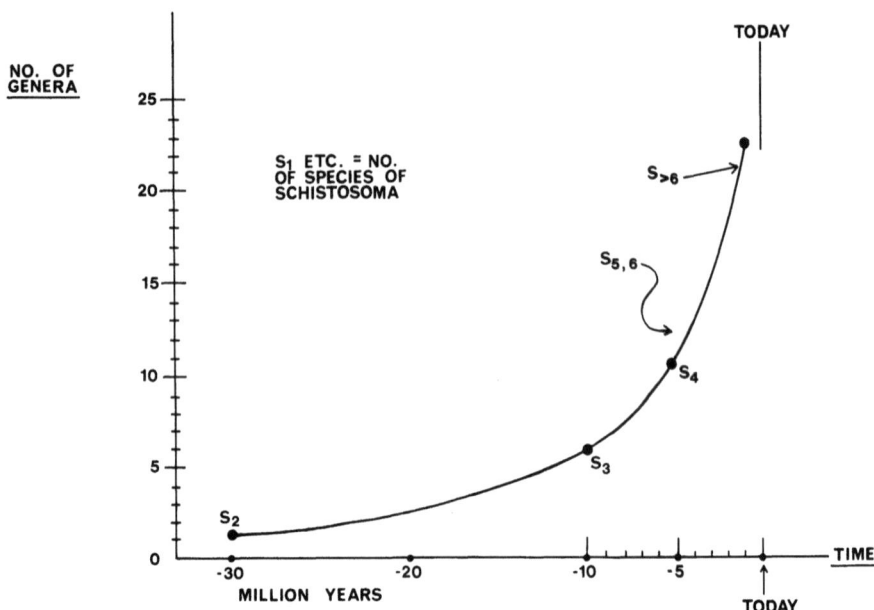

Figure 6.18. Hypothesis of the number of species of *Schistosoma* transmitted by triculines (1) identifiable at different moments in time based on reconstructions from cladograms; (2) relative to the number of triculine genera as they evolved and diverged; and (3) the six species *Schistosoma* known at present. Refer to Figure 6.19. Fossil evidence coupled with triculine genera now isolated in different drainages indicate that the following triculine genera (or their immediate precursors) had evolved at −10 m.y. to −7 m.y. *Paraprososthenia, Tricula, Neotricula, Karelainia, Hubendickia, Delavaya, Fenouilia, Kunmingia,* a precursor to *Robertsiella*.

S_2 = two species, pre-*japonicum*; pre-*sinensium*
S_3 = three species: (1) *japonicum*; (2) pre-*mekongi-malayensis*; (3) *sinensium*
S_4 = four species: (1) *japonicum*; (2) *mekongi*; (3) *malayensis*; (4) *sinensium*
$S_{5,6}$ = six species; the four above plus two additional, probably in the *sinensium* complex

ous studies spanning decades. The bar representing generic-level specificity (Fig. 6.20) extends to the present because *Pomatiopsis lapidaria* of North America can experimentally be infected with *S. japonicum*. The bar at the species (or subspecies) level also extends to the present because, as DeWitt (41) demonstrated, it is possible to infect snails of one region (China, Taiwan, Japan, Philippines) with a strain of *S. japonicum* from another, although not all combinations work. Most recently Yuan et al. (44) confirmed the findings of DeWitt by showing that *O. hupensis quadrasi* from the Philippines could not be infected with the Chinese parasite, although a low percentage (6.9%)

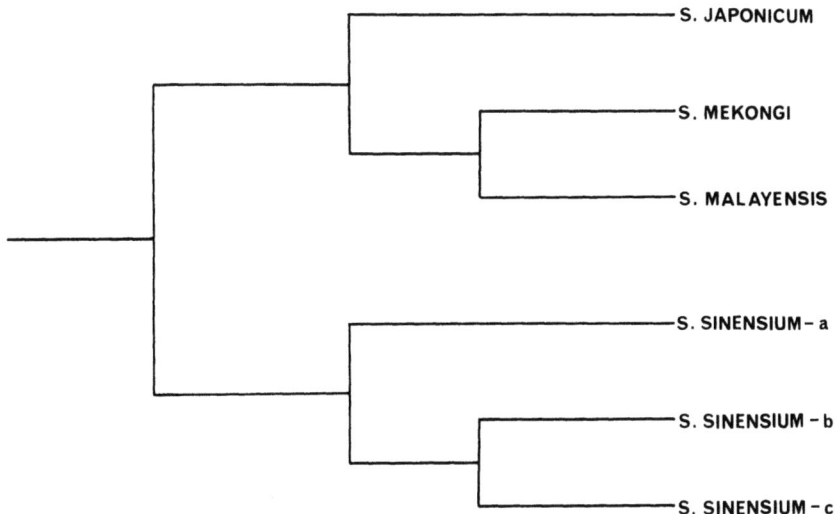

Figure 6.19. Hypothesis on the divergence of species of *Schistosoma*. The divergence of the *japonicum* complex of species, as predicted by Davis and Greer (6) on the basis of area and snail anatomy-based cladograms has been confirmed by use of molecular genetic data. (See text for details.)

of *O. hupensis hupensis* from China could be infected with the Philippine parasite (controls transmit at 34–49%). However, as discussed above, population specificity (below the subspecies level) has also been clearly demonstrated.

When considering the Triculinae, it is necessary to consider the *S. japonicum* complex apart from the *S. sinensium* complex (Table 6.4). In no case does *S. japonicum* develop in a triculine snail nor does *S. mekongi* or *S. malayensis* develop in *Oncomelania*, as shown by Yuan et al. (44) in laboratory infectivity studies. I interpret there facts to indicate divergence of these taxa earlier than divergence among triculine taxa and schistosomes associated with them when the Yangtze and Mekong River drainages separated. *S. mekongi* and *S. malayensis* show at least generic-level and probably species-level specificity (Table 6.5). However, *Tricula bollingi* from northwestern Thailand can be experimentally infected with *S. mekongi*, though with low level success. Thus for *S. mekongi*, generic-level specificity has not yet been fully attained and will probably not be attained for reasons discussed below.

The *Schistosoma sinensium* complex is apparently restricted to triculine snails of the tribes Triculini and Pachydrobiini. The *Schistosoma sinensium*-like parasite naturally infects *Tricula bollingi* and experimentally can infect *Neotricula aperta*. Here also, complete

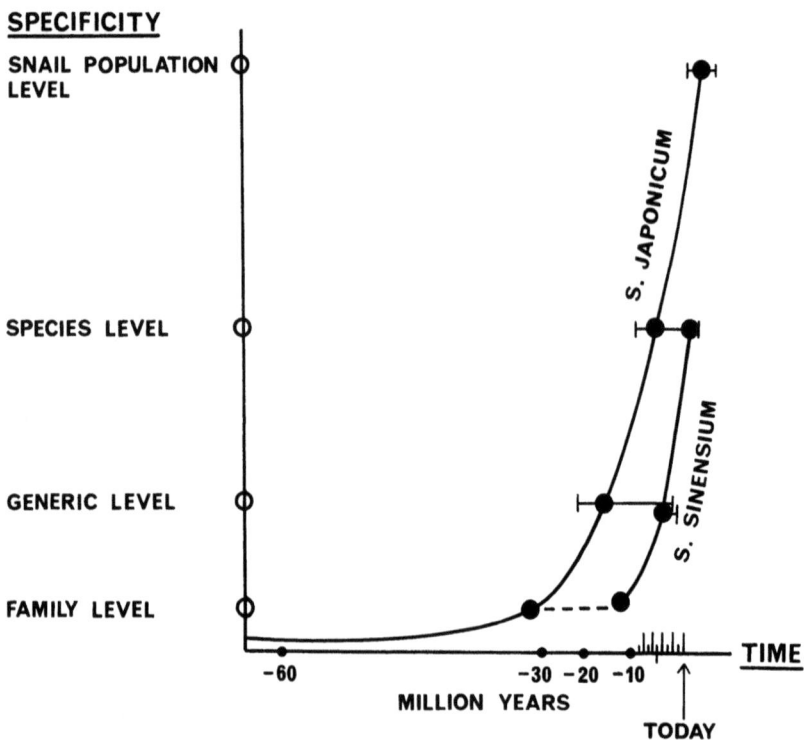

Figure 6.20. Hypothesis on the evolution of snail–schistosome coevolved specificity. See text for details.

generic-level specificity has not yet been attained. However, too little is known about S. sinensium sp. and Sinensium-like taxa to draw firm conclusions about specificity. I expect that future studies will uncover generic- and species-level coevolved specificities considering the overall geographic regions involved.

What deserves special attention is *Neotricula aperta*, now relictual in, and restricted to, the lower Mekong River with *S. mekongi*, and its potential to be infected with both *S. malayensis* and *S. sinensium* even if that potential is slight. Factors to focus on to explain these phenomena are (1) the time of divergence of river systems with triculine snails and *Schistosoma*; and (2) the ecology, population dynamics, and selective pressures operating on snails and parasites. *Neotricula* is hypothesized to have evolved from a *Tricula*-like ancestor (Fig. 6.11) initiating the Pachydrobiini clade of which *Robertsiella* is a more derived member. *S. japonicum* complex species capable of infecting man evolved in and are possibly restricted to this snail clade.

6. Evolution of Prosobranch Snails Transmitting Asian *Schistosoma* 191

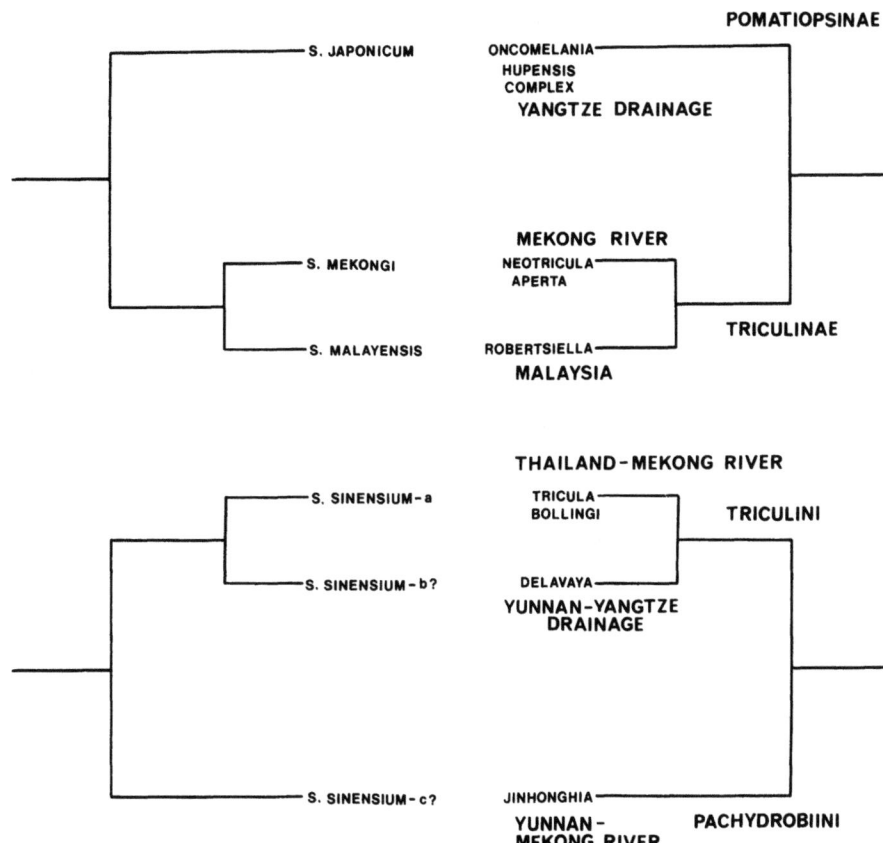

Figure 6.21. Testing congruence of cladograms. The snail cladograms are derived from Figures 6.10 and 6.11. The *S. japonicum* cladogram is based on congruence of area–river and molecular data cladograms. The relation among proposed *S. sinensium* species is hypothetical, but see text for details.

Table 6.5 Result of cross-infectivity studies involving four species of *Schistosoma* and four snail taxa.

Schistosoma taxa	Snail taxa			
	Oncomelania hupensis	Neotricula aperta	Robertsiella kaporensis	Tricula bollingi
Japonicum species group				
S. japonicum	+	−	−	−
S. mekongi	−	+	−	+ (L)
S. malayensis	−	+ (VL)	+	−
Sinensium species group				
S. sinensium (Thailand)	−	+ (VL)	−	+

Source: modified from Yuan et al. (44).
VL = very low infection rate; L = low infection rate.

The ecology of *Neotricula aperta* in the lower Mekong coupled with the ecology of transmission in the immense river is unique. Although *N. aperta* exists over a long stretch of river in populations that explode into millions each year owing to a single r-selected reproductive spurt, there are only few, tiny foci of infection in the down-river part of the snail distribution. No other population of *Tricula* or *Neotricula* is so extensive or reproduces in such numbers. In summary, we see a history of immense population size of an annual snail, channeled in one river system, with low infection rates in limited down-river foci. In such a system the genetic history of this snail species would not be much affected by *Schistosoma* transmission. The generalized open vulnerability (i.e., *Schistosoma*-naive) for *Schistosoma* transmission would be maintained at a high population level relative to ancestral experience with the ancestor to both *S. mekongi* and *S. malayensis*, and with *S. sinensium*. Without intense exposure to parasite infection in small populations (or at all) the open vulnerability to infection may be maintained indefinately (e.g., *O. h. chiui* and *Pomatiopsis lapideria* are such examples, discussed above). The hypothesis stands that infectivity pressure must be sufficient to obtain inheritable mutations in the snail that serve to deter the parasite. However, there has apparently been some impact due to historical contact with *S. "sinensium"* and the ancestor to *S. malayensis* and *S. mekongi* given the very low susceptibility levels of *Neotricula aperta* to *S. "sinensium"* and *S. malayensis* (Table 6.5).

Historical Reconstruction

Much as one can construct the evolution of the universe by studying fossil light from stars, it is reasonable to reconstruct historical events concerning the coevolution of pomatiopsid snails and Asian *Schistosoma* using data derived from geologic events, fossil evidence, systematic analyses with subsequent phylogenetic analysis, biogeographic analyses, and genetics. The facts must fit the model proposed; and as new data are obtained, all aspects of the model are challenged. The steps presented a decade ago still appear to be valid and have been strengthened with new data.

1. Fossil evidence from the Upper Cretaceous of India indicates an extensive marshy environment with few species of snails. The ancestral schistosome would have been associated nonspecifically with various snails of a given grade of organization, e.g., Prosobrachia: Pomatiopsidae.

2. With India colliding with the Asian plate causing the Himalayan orogeny, new environments were created. Snails invaded all freshwater environments. The advent of mountain streams and newly emerging drainage systems opened these new environments to the snails that became the aquatic Triculinae of today; one type of snail maintained the ecotone of an amphibious existence, becoming modern *Oncomelania* of the Pomatiopsidae. The attending nonspecific *Schistosoma* (two types: *japonicum*-complex and *sinensium*-complex precursors) moved into both habitats with these respective snail types.

3. Aquatic environments changed with extreme speed. Rapid episodic events of lake creation—destruction, stream evolution (with substratum evolution)—fractionated populations and drove rapid speciation. Schistosomes became divided ecologically. With fractionation of snail populations into numerous small populations came the advent of specificity. The schistosomes were always under one constraint during this and all subsequent periods: access to a definitive host; snails were not under such a constraint. As the snails rapidly evolved, the parasite had to keep pace. In small populations of snails, the parasite could exert considerable pressure relative to population survival. any mutation in the snail that could thwart the parasite would be of value [reviewed by Davis (1)]. A mutation that protected the snail from infection was countered by a mutation in the parasite to overcome the snail's defense. This genetic fencing match increased in complexity so that coevolving genetic trajectories in time and space diverged irrevocably from lineages now long isolated by evolving in an ecospace not available to schistosomes (e.g., Jullieniini, *Pachydrobia* of the Pachydrobiini, *Lacunopsis* of the Triculini) (Table 6.5). Coevolved specificity rapidly moved from the generalized family ground-plan arena to the species and then to the population level (Fig. 6.20). The cladograms of evolving snail lineages (Figs. 6.10 and 6.11) have marked on them the proved capability to transmit a species of *Schistosoma*. Also marked NS (*N*ot transmitting *S*chistosomes) are those lineages that, thus far, have not been shown to be able to transmit a schistosome. Notably ,the remarkable adaptive radiation of the Jullieniini involving specialized ecospace has not been shown capable of transmitting a schistosome (55) (Table 6.6). The clades involving taxa capable of transmitting a schistosome are unbroken, indicating a continuous genetic history of schistosome-transmitting potential, as one would predict from a hypothesis of coevolution toward increasing genetically involved specificity. The hypothesis would be seriously challenged if, for example, a species of *Hydrorissoia* or *Hubendickia* were found capable of transmitting a schistosome.

4. *Schistosoma japonicum* and *S. mekongi* ancestors would have

Table 6.6 Infectivity studies involving seven Mekong River snail taxa, eight snail genera, two species of *Schistosoma*.

Snail genera	S. mekongi	S. japonicum (5 strains)	S. japonicum Japenese strain	S. japonicum Hunan
Neotricula aperta[a,c] (3 races)	+	−	×	×
Pachydrobia bavayi[b]	−	×	−	×
Pachydrobia crocki[c]	−	×	×	×
Gammatricula shini[b,e]	−	×	−	×
Manningiella sp.[c]	−	×	×	×
Hubendickia sp.[c]	−	×	×	×
Lacunopsis sp.[c]	−	×	×	×
Stenothyra sp.[c]	−	×	×	×
Lithoglyphopsis modesta[d,f]	×	×	×	−

[a] Lian and Kitikoon (43)
[b] Lo et al. (56).
[c] Sornmani et al. (55).
[d] Chen Cui-E (1988, personal communication).
[e] Yonagunijima, off Tiawan.
[f] Human, China (type locality).
+ indicates successful infections; − indicates negative results; × indicates not attempted.

diverged with ecological separation and then separation of drainage systems some 12 ± 4 million years ago. Fletcher et al. (54) stated that, based on allozymic electrophoretic molecular genetics, an estimated Nei D of 1.71 to 2.40 between *S. mekongi* and *S. japonicum* suggested that these taxa diverged 8.5 to 12.0 million years ago. More recently, Yong et al. (53) showed that *S. malayensis* and *S. mekongi* were more similar to each other on the basis of allogymic molecular genetics (Nei D value of approximately 0.6) than either was to *S. japonicum*. These findings had been predicted on the basis of phylogenetic and area cladogram analysis (6). Subsequently, Woodruff et al. (57) provided a greater suite of allozyme data confirming that *S. mekongi* and *S. malayensis* have a Nei D value of 0.6, and these taxa are highly divergent from *S. japonicum*. Given a calibrated molecular clock, the 0.6 index indicates a divergence of some 4 million to 5 million years ago (Figs. 6.18 and 6.21). Note the match of cladograms for species of the *S. japonicum* complex and the divergence of relevant snail taxa (Fig. 6.20).

Two *Schistosoma* species groups have evolved in parallel in the more generalized Triculine snails (Fig. 6.19). Much work is necessary with the *S. sinensium* complex in China before more can be added to the discussion above. A hypothesis of evolved relation among relevant snail taxa and taxa of the *S. sinensium* complex is given in Figure 6.21 (bottom).

Ecology of Transmission

Adaptive and Morphostatic Radiations

Davis (7,46) extensively reviewed the concept of the Triculinae adaptive radiation. Osborn (58) first used the term "adaptive radiation," and Simpson (59) fully discussed the concept: "Adaptive radiation is, descriptively, this often extreme diversification of a group [e.g., mammalian or reptilian radiations] as it evolves in all the different directions permitted by its own potentialities and by the environments it encounters." Stanley (60) stated: "Adaptive radiation is the rapid progression of new taxa from a single ancestral group."

There are two themes set forth by Osborn and Stanley, and they are not necessarily concordant: (1) extreme diversification and (2) rapid progression [proliferation?] of new taxa in monophyly. The Triculinae have done both but not both for all clades. The Jullieniini, *Pachydrobia* of the Pachydrobiini, and *Lacunopsis* of the Triculini exhibit extreme morphologic and ecological diversification as well as prolific speciation. I restrict the term adaptive radiation to this phenomenon. The dimensions of diversification are (1) reproductive organ changes associated with reproduction; (2) radular changes associated with feeding diversification; (3) shell shape and sculptural changes associated with habitat diversification and (speculation) species recognition when many congeners are in sympatry (e.g., *Hubendickia*). The exuberant exploitation of every conceivable niche by this splendid adaptive radiation in the lower Mekong has been documented (5,7,36).

A rapid proliferation of monophyletic taxa has also occurred that does not exhibit extreme anatomical or ecological diversification. It is seen in most Triculini and Pachydrobiini. With dispersal down evolving river systems, considerable speciation has occurred in isolated hilly habitats with tiny perennial cold, clear-water streams. The shells of these species have not diversified much nor has the general ground plans of the reproductive or radular systems. The situation of considerable, rapid speciation with low anatomical diversification I here call "morphostatic radiation." In this type of radiation there is little or no habitat diversification, speciation is widely allopatric, and there are low levels of anatomical change. Thus we find *Tricula*, *Neotricula* and *Gammatricula* have virtually identical type shells (Fig. 6.7), male reproductive systems, and radula.

Together, the three triculine clades that exhibit adaptive and morphostatic radiations may be considered a macroadaptive radiation (46) in that when considering higher taxa there is a gradation from the spectacular adaptive radiation in the lower Mekong River to the taxa that demonstrate morphologic diversity that supports the definitions of more then 20 genera. Parts of some clades exhibit the pattern of a morphostatic radiation so common throughout China. The adaptive and morphostatic radiations of the macroadaptive radiation have been further divided (46): A null radiation is a monotypic genus, a taxon recognized by the discrete morphologic gap from the other genera to which it is phyletically allied. Such a genus may be the basis for a future first order radiation, or it was represent a dead end. *Saduniella*, with its planispiral shell, of the Jullieniini is such a genus. A first order radiation involves a single genus with at least two, but usually more, species, e.g., *Pachydrobia, Tricula, Hubendickia*, A second order radiation involves two or more first order radiations. The tribe Jullieniini comprises a second order adaptive radiation.

Ecology of Transmission

General habitat types of snail hosts of Asian schistosomes are given in Table 6.7. *Oncomelania* lives from the mountain highlands in Yunnan, China to marshy lowlands in Hunan. It is sympatric with species of *Tricula* in Yunnan as well as Zhejiang Province. However, *Oncomelania* is highly amphibious. *Tricula* is aquatic but there is considerable variance. Some species of *Tricula*, though primarily aquatic, demonstrate capability for leading an amphibious existence, at least for long periods. In Yunnan, *Oncomelania* is sympatric with *Tricula gregoriana* (possibly also with other species).

Tricula inhabits tiny mountain streams that are mostly a meter or less wide with a depth of a few centimeters (61). The habitat is shaded and the water cool and clear, frequently arising from a spring. One species, *T. xiaolongmenensis* from Yunnan, lives in microsympatry with *Pseudobythinella* under rocks embedded in a cliff; the rocks are wet, and the snails are not in running water. Species of *Tricula* observed to be infected with *Schistosoma* live in the tiny mountain streams. The species of this genus primarily live under rocks, leaves, and sticks on the stream bed. *Jinhongia* lives as does *Tricula*.

Neotricula lives in a wider variety of habitats than does *Tricula* ss. While most species of *Neotricula* live as do species of *Tricula*, the host of *Schistosoma mekongi* (*Neotricula aperta*) lives in the lower Mekong River in relatively quiet flowing or still water (during the dry season) in a few centimeters of water on aquatic plants, roots, and sticks. The species is sympatric with species of *Hubendickia, Hydrorissoia*, and *Stenothyra*. In Guangxi Province a triculine (*Tricula* or *Neotricula*, species not yet determined) lives in a narrow irrigation ditch in a rice field close to the source of water, a spring of cold clear water; it lives in sympatry with a species of *Stenothyra*. Another lives

Table 6.7 General habitats of species of schistosome transmitting and nontransmitting pomatiopsids.

Species	Mountain streamlet	Jungle stream	Spring-fed pool	Lake bottom	Mountain seepage	Irrigation canals marshes, streamlets: amphibious	River
Transmitters							
Oncomelania hupensis	−	−	−	−	−	+	−
Delavaya dianchiensis	−	−	+	−	−	−	−
Tricula bollingi	+	−	−	−	−	−	−
Neotricula aperta	−	−	−	−	−	−	+
Jinhongia jinghongensis	+	−	−	−	−	−	−
Robertsiella kaporensis	−	+	−	−	−	−	−
Nontransmitters							
Tricula montana	+	−	−	−	−	−	−
Tricula spp.	+	−	+	−	+	−	−
Lithoglyphopsis	−	−	−	−	−	−	+
Lacunopsis	−	−	−	−	−	−	+
Fenouilia	−	−	−	+	−	−	−
Neotricula spp.	+	−	+	−	−	−	−
Halewisia	−	−	−	−	−	−	+
Pachydrobia	−	−	−	−	−	−	+
Wuconchona	−	−	−	−	+	−	−
Gammatricula	+	−	−	−	−	−	−
Kunmingia	−	−	+	−	−	−	−
Other Jullieniini	−	−	−	−	−	−	+

in a wide stream some 2 meters wide and 3–6 cm deep, under rocks, in sympatry with a species of *Assiminea*.

Robertsiella lives in jungle streams 0.3–4.5 meters wide and 0.9–15.0 mm deep flowing over sandy-gravel or sandy-mud bottoms. The snails live among the roots of the *Saraca* trees. These trees send thick mats of roots below the surface of the water. Sympatric with *Robertsiella* are *Brotia costula*, *Melanoides tuberculata*, and *Thiara scabra* (all Thiaridae). The habitat of *Robertsiella* is close to the Jelai River, which is large enough to navigate by small boat. Thus both species of the Pachydrobiini clade that transmit *Schistosoma* to man live in or are associated with a river environment.

Those taxa that entered new ecospace and underwent adaptive radiation do not transmit *Schistosoma*. *Lacunopsis* (Fig. 6.8 and 6.9), a large first order radiation, is a dominant faunal element in the lower Mekong River. The niche dimensions filled are swift-water habitats on rocks where species differences are seen in shell shape, size, and sculpture associated with positional relations in the water column involving rock slope, depth, surface, and degree of current. Shell shapes (Figs. 6.8 and 6.9) are astonishing for freshwater gastropods as shapes converge on those of marine Neritidae, Littorinidae, and Fossaridae (7).

Pachydrobia, closely related to *Neotricula* (Pachydrobiini) has also undergone a splendid adaptive radiation in the lower Mekong. Species invaded a variety of habitats especially in muddy sand and sand; one species specialized in living in shallow pools on rock islands exposed during the dry season when river levels drop 20–30 feet. Shell shape and sculpture vary considerably. The shell is asymmetric owing to the flattened face of the body whorl. Shells are thick; and they may be smooth (Fig. 6.4t,u), finely ribbed, with heavy nodular ribs (Figs. 6.4r and 6.8h), irregular bosses, or spines. Sizes (shell lengths) range from 5 to 14 mm, and in *Lancunopsis* the range is from 6.2 to 18.0 mm. Contrast these figures with those for species of *Tricula*, where sizes range from about 2.8 to 5.6 mm, a span of only 2.8 mm.

The Jullieniini comprise a spectacular second order adaptive radiation in the lower Mekong River with five first order radiations and two null radiations. *Karelainia* is the bridge between the Triculini and the Jullieniini, the least modified anatomically compared to *Tricula bollingi* or *Tricula montana*. The gonadal morphologies are the same, and the oviduct coil is almost as narrow as the 360° closed twist in the Triculini (Figs. 6.3 and 6.15). However, the average length of shells of *Karelainia* (6.1 mm; $n = 4$) is one third again the size of the average *Tricula* or *Neotricula* (4.0 mm, $n = 13$). Excluding *Karelainia*, the second order adaptive radiations average 13.5 species per genus (range 9–17; $n = 6$) in the lower Mekong River (5). Why then does *Karelainia* have only four species? On the basis of shell size, shape, and sculpture (smooth, one or two spinal rows of cords, several spiral rows of cords, or fluted spines) species of *Karelainia* and *Neoprososthenia* so much resemble each other that they were previously all described as *Paraprososthenia* [now = *Neoprososthenia*]. They live in similar ecologic conditions, i.e., slight to moderate current on rocks with silty sand or sandy mud. Two or more species of *Neoprososthenia* are usually sympatric with a species of *Karelainia*. Together these tow genera fill out a radiation (14 species) equal to that of *Hubendickia* or *Pachydrobia*. Not all species currently classified as *Neoprososthenia* have been studied anatomically; thus it is possible that some species might yet be removed from *Neoprososthenia* and be classified as *Karelainia*.

Hubendickia deserves considerable attention. It is a large first order radiation of no fewer than 19 species—no fewer than 16 in the lower Mekong River. It, of all Jullieniini, has the greatest distribution (Nan Pan–Pearl drainage system in Yunnan; Mekong of Yunnan to below the fall line in Cambodia). Considering shell characteristics and tooth morphology coupled with transition states of seminal receptacle size, we see a genus that represents conditions that are close to basal in the Jullieniini. (If one coupled the above characteristics with *Karelainia*'s male and female reproductive system, one would see "the" ancestor!) The shell ranges from the primitive thin smooth, small, ovate-conic type to fusiform to turreted. Sculpture increases to ribs and reticulate. The central tooth of the radula varies from the general Triculini type [type I of Davis (5)] (Fig. 6.15) to derived types. Only *Hubendickia* of the Jullieniini genera has species with a considerable range of central tooth morphology that includes the generalized type. In one species the central cusp is flattened and broadened for scraping in mud; in other species the anterior central cusp support is massive and thickened, apparently adapted to reduce wear while scraping and shreading material from hard substrata. It is in *Hubendickia* that there is an indication that certain sculptured character-states are related to species recognition when species live in sympatry; we see a possible case for character displacement (at Khemarat, Thailand, five species live sympatrically). It is common to find hundreds each of four species in a handful of algae; and each species has a distinctive sculpture involving ribs. Sculpture degrades downstream when these species are not sympatric (7).

Jullienia deserves comment as this primary adaptive radiation of some 16 species has radiated, as did *Lacunopsis*, in moderate to heavy current in rocky habitats. It has highly derived reproductive, radular, and shell character-states.

All Triculinae are semelparous. All live only 1 year or slightly longer. Once *Pachydrobia* reproduces, the reproductive system slowly disintegrates, which is first seen in the male where the penis begins to disintegrate. It is seen later in the female where the ovary and pallial oviduct disintegrate. The snails live on for a month or more after the onset of disintegration occurs. There is evidence that this phenomenon occurs in other genera (5,46). Once *Neotricula aperta* has laid eggs, it dies; and there is a period of about 1 month when no adults are seen and no hatched young can be found.

In the Mekong River, growth and reproduction are in phase with the annual river cycles. Different groups of species mature, reproduce, and die at different times once the dry season begins and water levels start to drop in October. Habitats then begin to emerge and form. First island masses and the larger waterfalls appear, followed by smaller islands, embayments between islands, sand bars, lakes, pools on islands, smaller rapids, and finally shallow, quiet areas that allow for considerable mud deposition. From mid-October or November through June most habitats are free from flooding and destruction caused by monsoons (7). Each year old habitats are destroyed and new ones created. Over the two million plus years since the Yangtze and Mekong Rivers were close to being totally isolated, the Mekong River has cut down through the Korat Plateau creating new island-rapids environments ever further downstream and filling in behind with micaceous sand. Now *Schistosoma mekongi*, transmitted by only the single most generalized species found in the lower Mekong River, and the macroadaptive radiations of the lower Mekong straddle the fall line and are on the verge of being swept to extinction in brackish water within a short span of geologic time.

Acknowledgment. This review is dedicated to my colleagues in China without those help this work could not be accomplished: Drs. Chen Cui-É, Guo Y.H., Kang Z.B., and Liu Y.Y. Photographs, illustrations, and graphics were done by Caryl Hesterman, Elizabeth Carroza, and Susan Trammell. The work is supported by NIH grant AI 11373.

References

1. Davis GM: Snail hosts of Asian *Schistomosoma* infecting man: origin and coevolution. *In:* Bruce, et al. (eds.), *The Mekong Schistosome*, Malacological Review, Supplement 2; 1980;195–238.
2. Annandale N: The molluscan hosts of the human blood fluke in China and Japan, and species liable to be confused with them. In Faust, EC and Meleney, HE (Eds.), Studies on schistosomiasis japonica. American Journal of Hygiene Monographic Series No. 3, 1924;269–294, 1 pl.

3. Annandale N & Prashad, B: Contribution to the fauna of Yunnan based on collections made by J Coggin Brown, BSc, 1909-1910. Part IX. Two remarkable genera of freshwater gastropod molluscs from the Lake Erh-Hai. Records of the Indian Museum 1919;16:413-423.
4. Taylor DT: Summary of North American Blancan non-marine mollusks. Malacologia 1966;4(1):1-172.
5. Davis GM: The origin and evolution of the gastropod family Pomatiopsidae, with emphasis on the Mekong River Triculinae. Monograph of the Academy of Natural Sciences of Philadelphia No. 20, 1979:1-120.
6. Davis GM & Greer, G: A new genus and two new species of Triculinae (Gastropoda: Prosobranchia) and the transmission of a Malaysian mammalian *Schistosoma* sp. Proceedings of the Academy of Natural Sciences of Philadelphia 1980;132:245-276.
7. Davis GM: Different modes of evolution and adaptive radiation in the Pomatiopsidae (Prosobranchia: Mesogastropoda). Malacologia 1981; 21:209-262.
8. Davis GM, Kuo YH, Hoagland KE, Chen PL, Yang HM, & Chen DJ: Advances in the Systematics of the Triculinae (Gastropoda: Prosobranchia): the genus *Fenouilia* of Yunnan, China. Proceedings of the Academy of Natural Sciences of Philadelphia 1983;135:177-199.
9. Davis GM, Kuo YH, Hoagland KE, Chen PL, Yang HM, & Chen DJ: *Kunmingia*, a new genus of Triculinae (Gastroposa: Pomatiopsidae) from China: Phenetic and cladistic relationships. Proceeding of the Academy of Natural Sciences of Philadelphia 1984;136:165-193.
10. Davis GM, Liu YY, Chen, YG: New Genus of Triculinae (Prosobranchia: Pomatiopsidae) from China: Phylogenetic Relationships. Proceedings of the Academy of Natural Sciences of Philadelphia 1990;142:143-165.
11. Gredler PV: Zur Concylien-Fauna von China. Jahrbücher der Deutschen Malakozoologischen Gesellschaft 1881;8:119-132.
12. Liu YY, & Zhang YZ: On new genus and species of freshwater snails harbouring cercariae of lung flukes from China. Acta Zootaxonomica Sinica 1979;4(2):132-136.
13. Kuroda T, & Habe T: New aquatic gastropods from Japan. Venus 1954;18(2):71-78.
14. Brandt, RAM: Decriptions of new non-marine mollusks from Asia. Archiv für Molluskenkunde 1968;98:213-289.
15. Bavay, A: Conquilles novelles, provenant de récoltes de ML Levay, dans les râpides. du Haut-Mékong, pendant la campagne du Massie. Jorunal de Conchyliologie 1895;43:82-94, pls. 5-6.
16. Crosse H, & Fischer P: Mollusques fluviatiles, rucueillis au Cambodge, par la mission scientifique francaise de 1873. J Conchyl 1876;24:313-342, pls. 10, 11.
17. Thiele J: Revision des systems der Hydrobiiden und Melaniiden. Abt System Okol Geogr Tiere 1928;55:351-402.
18. Annandale N: The gastropod fauna of old lake beds in upper Burma: Records of the Geological Survey of India 1919;50:209-240.
19. Brandt RAM: New freshwater gastropods from the Mekong. Arch Mollusk 1970;100:183-205.

20. Davis GM, & Zaibin Kang: The genus *Wuconchona* of China (Gastropoda: Pomatiopsidae: Triculinae): anatomy, systematics, cladistics, and transmission of *Schistosoma*. Proc Acad Nat Sci Phila 1990;142:119–142.
21. Davis GM, Guo YH, Hoagland KE, et al: Anatomy and systematics of triculine (Prosobranchia: Pomatiopsidae: Triculinae) freshwater snails from Yunnan, China, with descriptions of new species. Proc Acad Nat Sci Phila 1986;138:466–575.
22. Kang ZB: A new genus and three new speices of the family Hydrobiidae (Gastropoda: Prosobranchia) from Hubei Province, China. Oceanol Limnol Sinica 1983;14:499–505.
23. Heude RPM: Diagnoses Molluscorum novarum in Sinis collectorum. J Conchyl 1889;37:40–50.
24. Deshayes PG, & Jullien J: Mémoire sur les mollusques nouveaux du Cambodge envoyés au museum par M. le docteur Jullien. Bull Nouv Arch Museum 1876;10:115–162.
25. Benson WH: Description of *Camptoceras*, a new genus of the Lymnaeidae, allied to *Ancylus*, and of *Tricula*, a new type of form allied to *Melania*. Calcutta J Nat Hist 1843;3(12):465–468.
26. Malek EA: *Snail Hosts of Schistosomiasis and Other Snail-transmitted Diseases in Tropical America: A Manual*. Pan American Health Organization. Washington, DC, 1985. 325 pp.
27. Davis GM, Kuo YH, Hoagland KE, et al: *Erhaia*, a new genus and new species of Pomatiopsidae from China (Gastroposa: Rissoacea). Proc Acad Nat Sci Phila 1985;137:48–78.
28. Ioganzen BG, & Starobogatov YI: A finding of a freshwater mollusc of the Family Triculidae (Gastropoda, Prosobranchia) in Siberia. Zool Zh 1982;61:1141–1147.
29. Tryon GW: Notes on American fresh water shells, with descriptions of two new species. Proc Acad Nat Sci Phila 1862;14:451–452.
30. Say T: Description of seven species of American fresh water and land snails, not noticed in the systems. J Phila Acad Sci 1817;1:13.
31. Yen TC: Die chinesischen Land-und Susswasser—Gastropoden des Naturmuseums senckenberg. Abh Senckenbergisch Naturforasch Ges 1939;444:1–234.
32. Zilch A: Vinzenz Gredler und die Erforschung der Weichtiere Chinas durch Franzikaner aus Tirol. Arch Mollusk, 1974;105:171–228.
33. Liu YY, Wang YX, & Zhang WZ: On new species and new records of freshwater snails of the family Hydrobiidae from Yunnan, China. Acta Zootaxon Sinica 1980;5:358–368.
34. Wilson EO: A consistency test for phylogenies based on contemporaneous species. Syst Zool 1965;14:214–220.
35. Davis GM: New Tricula from Thailand. Arch Mollusk 1968;98:291–317.
36. Davis GM: Historical and ecological factors in the evolution, adaptive radiation, and biogeography of freshwater mollusks. Am Zool 1982;22:375–395.
37. Ponder W: The truncatelloidean (=rissoacen) radiation—a preliminary phlogeny. Malacol Rev [Suppl 4] 1988:129–164.
38. Davis GM, Rao NVS, & Hoagland KE: In search of Tricula (Gastropoda:

Prosobranchia): Tricula defined, and a new genus described. Proc Acad Nat Sci 1986;138:426–442.
39. Janzen DH: When is it coevolution? Evolution 1980;34:611–612.
40. Greer GJ, Ow-Yang CK, & Yong HS: *Schistosoma malayensis* n. sp.: *Schistosoma japonicum*-complex schistosome from penninsular Malaysia. J Parasitol 1988;74:471–480.
41. DeWitt WB: Susceptibility of small vectors to geographic strains of Schistosoma jaonicum. J Parasitol 1954;40:453–456.
42. Davis GM: *Oncomelania* and the transmission of *Schistosoma japonicum*: a brief review. In Harinasuta C (ed): Proceedings of the Fourth Asian Seminar on Parasitology and Tropical Medicine, Schistosomiasis and Other Snail-Transmitted Helminthiasis, Manila. 1969, pp. 93–103.
43. Liang YS, & Kitikoon V: Susceptibility of *Lithoglyphopsis aperta* to *Schistosoma mekongi* and *Schsitosoma japonicum*. Malacol Rev [Suppl] 1980;2:53–60.
44. Yuan HC, Upatham ES, Kruatrachue M, & Khunborivan V: Susceptibility of snail vectors to oriental anthropophilic *Schistosoma*. Southeast Asian J Trop Med Public Health 1984;15:86–94.
45. Cain AJ, & Harrison GA: Phyletic weighting. Proc Zool Soci Lond 1960;135:1–31.
46. Davis GM: Introduction to the second international symposium on evolution and adaptive radiation of Mollusca. Malacologia 1981;21:1–4.
47. Cain AJ: The pefection of animals. In Carthy JD, Duddington CL (eds): Viewpoints in Biology, No. 3. Butterworths, London, 1964; pp. 36–63.
48. Pao TC: The description of a new schistosome *Schistosoma sinensium* sp. nov. (Trematoda: Schistosomatidae) from Szechuan Province. Chinese Med J 1959;78:278.
49. Baidikul V, Uptham ES, Kruatrachue M, et al: Study on *Schistosoma sinensium* in Fang District, Chiangmai Province, Thailand. Southeast Asian J Trop Medi Public Health 1984;15:141–147.
50. Sun CC: Notes on some *Tricula* snails from Yunnan Province. Acta Zool Sinica 1959;11:460.
51. Kruatrachue M, Upatham EC, Sahaphong S, et al: Scanning electron microscopic study of the tegumental surface of adult *Schistosoma sinensium*. Southeast Asian J Trop Med Public Health 1983;14:427–438.
52. Greer GJ, Kitikoon V, & Lohacit C: Morphology and life cycle of *Schistosoma sinensium* Pao, 1959, from northwest Thailand. J Parasitol 1989; 75:98–101.
53. Yong HS, Greer GJ, & Ow-Yang CK: Genetic diversity and differentiation of four taxa of Asiatic blood flukes (Trematoda, Scistosomatidae). Trop Biomed 1985;2:17–23.
54. Fletcher M, Woodruff DS, LoVerde PT, & Ash HL: Genetic differentiation between *Schistosoma mekongi* and *S. japonicum*: an electrophoretic study. Malacol Rev [Suppl] 1980;2:113–122.
55. Sornmani S, Kitikoon V, Schneider CR, et al: Mekong schistosomiasis. 1. Life cycle of *Schistosoma japonicum*, Mekong strain in the laboratory. Southeast Asian J Trop Med Public Health 1973;4:218–225.
56. Lo CT, Berry EG, & Iijima T: Studies on schistosomiasis in the Mekong

Basin. II. Malacological investigations on human Schistosoma from Laos. Chinese J Microbiol 1971;4:168–181.
57. Woodruff DS, Merenlender AM, Upatham SE, & Viyanant V: Genetic variation and differentiation of three Schistosoma species from the Phillippines, Laos and peninsular Malaysia. Am J Trop Med Hyg 1987; 36:345–354.
58. Osborn HF: The Origin and Evolution of Life. Scribner's Sons, New York, 1918. 322 pp.
59. Simpson GG: The Meaning of Evolution. Yale University Press, New Haven, CT, 1949. 322 pp.
60. Stanley SM: Macroevolution. WH Freeman, San Francisco, 1979. 364 pp.
61. Hoagland KE, Kuo YH, Davis GM, et al: Ecology and zoogeography of some mainland Chinese Tricula (Gastropoda: Probranchia: Pomatiopsidae) transmitting schistosomes. Am Malacol Bull 1984;2:88.
62. Raven PH, & Axelrod DI: Plate tectonics and Australian paleobiogeography. Science 1972;176:1379–1386.
63. Davis GM, Chen CE, Wu C, Kuang TF, Xing YG, Li L, Liu WJ, & Yan YL: The Pomatiopsidae of Hunan, China (Gastropoda: Rissoacea. Malacologia 1992;34:143–342.
64. Farris J, Hennig 86, Version 1.5. Stony Brook, N.Y. [Computer Software]. 1989.
65. Temcharoen P: New aquatic mollusks from Laos. Arch Mollusk 1971;102:91–109.
66. Voge M, Bruckner D, & Bruce J: *Schistosoma mekongi* sp. nov. from man and animals, compared with four geographic strains of *Schistosome japonicum*. J. Parasit. 1978;64:577–584.
67. Katsurada F: *Schistosoma japonicum*, a new parasite of man, by which an endemic disease in various areas of Japan is caused. Annot. Zoolgy of Japan 1904;5:146–160 [in German].

Index

A
Abscess-granuloma stage, 75, 80, 82, 83
Abscess stage, 74–75, 80
Adaptive radiation, 195–196
AGDR epitope, 110–111
AGML, *see* Gastric mucosal lesion, acute
AIDS
  cryptosporidiosis and, 9–10, 11–12
  toxoplasmosis and, 27, 28
Amastigotes, 119
Anchusan, 65
Anergic diffuse cutaneous leishmaniasis (DL), 120
Anethole, 65
Anisakiasis, *see* Anisakidosis
Anisakidae, 49
Anisakidosis, vi, 43–94
  acute gastric, 68–69, 72–73
  acute intestinal, 69–71, 77–79
  chronic gastric, 69, 74–77
  chronic intestinal, 71–72, 79–83
  clinical aspects, 56–64
  ectopic, 56
  epidemiologic aspects, 45–49
  extragastrointestinal, 64
  gastric, 45
  heterologous, 56
  histologic findings, 72–84
  immunology, 84–87
  intermediate or paratenic host as
    source of, 47, 49
  intestinal, 45
  intraperitoneal, 61, 64
  life cycle of parasites, 51–56
  location of disease, 66–68
  macroscopic findings, 68–72
  morphology of parasites, 50
  occurrence in world, 44
  pathologic classification, 66
  pathology, 65–84
  prevention, 64–65
  serodiagnosis, 87–90
  simultaneous multiple infections, 46–47, 48
  symptoms, 56
  taxonomy of parasites, 49–50
  tingling throat, 46
Anisakis genera, 49–50
Anisakis larvae, 76–78
  molecular genetics of, 90–93
Anisakis Physeteris, life cycle of, 53–54
Anisakis simplex
  life cycle of, 51–53
  occurrence in world, 44
  third-stage larvae, 53
Anisakis simplex larvae
  monoclonal antibodies against, 84–87
Antibody induction with subunit vaccines, 108–109
Antibody levels, longevity of, 112

205

Index

Antibody-mediated immunity, 110–111
Antibody-mediated passive immunization, 112–113
Aprinocid-N-oxide, 37
Area cladograms, 174, 175, 187
Ascites, 70
Asia, major rivers of, 149
Auerbach plexus, ganglion cells in, 79, 81
Autapomorphies, 170–171
Azithromycin, 12–13, 36, 37
AZT (zidovudine), 12

B
Bovine transfer factor, 12
Bursa copulatrix, 179

C
Charcot-Leyden crystals, 8
China, Triculini and Pachydrobiini in, 181
Chorioretinitis, 26
Circumsporozoite (CS) protein, 105–111
CL, see Cutaneous leishmaniasis
Cladograms, 168, 169, 173
  area, 174, 175, 187
  testing congruence of, 191
Clarithromycin, 36
Clindamycin, 35
Coevolution, 182
Colony-stimulating factors, 136
Congruence of cladograms, testing, 191
Contracaecum osculatum, 44
Contributors, xiii–xiv
Convergence, 184–185
Copepods, 54–55
Cross-infectivity studies, *Schistosoma*, 191
Cryptosporidial enteritis, 10
Cryptosporidiosis, v, 1–15
  AIDS and, 9–10, 11–12
  animal models, 13
  clinical presentation, 7–9
  diagnosis, 10–11
  epidemiology, 5–6
  fulminant, 9–10, 11–12
  human clinical manifestations, 8
  in vitro infection, 13–15
  microbiology, 2–5
  transmission, 6–7
  treatment, 11–13
Cryptosporidium, v, 1–8
  life cycle of, 3
  nosocimial spread of, 7
Cryptosporidium parvum, 2
CS (circumsporozoite) protein, 105–111
Cutaneous leishmaniasis (CL), 119
  anergic diffuse (DL), 120
Cytokines, 133
  in leishmaniasis, 132–136

D
Dapsone, 36
Delavaya, 146
DHFR (dihydrofolate reductase inhibitor), 35
DHPS (dihydropteroate synthetase), 35
Diarrhea, 2
  traveler's, 6
Diclazuril sodium, 12
Diffuse cutaneous leishmaniasis, anergic (DL), 120
Dihydrofolate reductase inhibitor (DHFR), 35
Dihydropteroate synthetase (DHPS), 35

E
Ectopic anisakidosis, 56
Editorial Board, xi
Encephalitis, toxoplasmic, 21, 31, 33–36
Enteritis, cryptosporidial, 10
Epitope specificity, 110–111
Erhaia, 162–163
ES (excretory-secretory) antigen, 84
Euphausiids, 52–53
Eusutoma rotundatum, 44
"Excervation" reaction, 75
Excretory-secretory (ES) antigen, 84
Exoerythrocytic phase, 103
Extragastrointestinal anisakidosis, 64

F
Fab fragments, 105–106, 113
Fat malabsorption, 8
Fenouilia, 146
Folinic acid, 35

G

Gamma interferon, 24, 37, 123, 133–135
Ganglion cells in Auerbach plexus, 79, 81
Gastric anisakidosis, 45
Gastric mucosal lesion, acute (AGML), 61
  etiology, 62
  relationship between anisakidosis and vanishing tumor, 63
Gastric pseudoterranovosis, 45–46
Gastroscopy, 45
GM-CSF (granulocyte macrophage colony stimulating factor), 124
Gondwanaland, 148
gp30 antigen, 131
gp42 antigen, 131
gp63 antigen, 129–131
Granulocyte macrophage colony stimulating factor (GM-CSF), 124
Granuloma stage, 75, 77, 83

H

Heiisan, 65
Heterologous anisakidosis, 56
HT29 cells, 14
Hubendickia, 199
Human immunodeficiency virus, 2
Hydrobiidae, 157
Hydroxynapthaquinone 566C80, 37
Hyperimmune cow colostrum, 12

I

IFN, *see* Interferon
IgG (immunoglobulin G) antibodies, 30
Ikasashi, 64–65
IL-2, 136
Immunity, antibody-mediated, 110–111
Immunization, antibody-mediated passive, 112–113
Immunoglobulin G (IgG) antibodies, 30
Infectivity studies, *Schistosoma*, 194
Interbelurin 2, 37
Interferon, gamma, 24, 37, 123, 133–135
Intestinal anisakidosis, 45
Intestinal pseudoterranovosis, 46
Intraperitoneal anisakidosis, 61, 64

J

Jullienia, 199
Jullieniini, 198
Jullieniini Davis, 164–165

K

Karelainia, 198–199

L

Lactose intolerance, 8
Lacunopsis, 198
Leishmania antigens, 122
  stimulating human T-cells, 128–132
Leishmania complex, 119
Leishmania immunology, 120
Leishmaniasis, 119
  acute visceral (VL), 120
  cellular responses in, 120–124
  cutaneous, *see* Cutaneous leishmaniasis
  cytokines in, 132–136
  delayed hypersensitivity responses, 120–121
  in vitro lymphocyte responses, 121–123
  lesions of, 126
  lymphocyte responses in, 123–124
  mucosal (ML), 119–120
  T-cell responses in, vi–vii 119–137
  T-cell subsets in human, 125–128
Letrazuril, 12
Lipophosphoglycan (LPG), 131–132
Lithoglyphopsis, 146
LPG (lipophosphoglycan), 131–132
Lymph nodes, mesenteric, 71
Lymphadenitis, toxoplasmic, 31
Lymphadenopathy, 26
Lymphocyte responses, in leishmaniasis, 123–124

M

Macrogametocytes, 5
Macrolides, 36
Malaria sporozoites, vi, 103–107
Malarial vaccines, vi, 103–114
Marsden, Philip Davis, vii–viii
Merozoites, 3, 5
Mesenteric lymph nodes, 71

Microgametocytes, 5
ML (mucosal leishmaniasis), 119–120
Monoclonal antibodies, against *Anisakis simplex* larvae, 84–87
Monophyly, 184
"Montenegro" test, 121
Morphology, 165, 171, 173
Morphostatic radiation, 195–196
Mucosal leishmaniasis (ML), 119–120

N
Neotricula aperta, 190, 192

O
Ocular toxoplasmosis, 26
Oncomelania hupensis paradigm, 183–184
Oocyst shedding, 5, 9
Oocysts, 5

P
p23 antigen, 37
Pachydrobia, 198
Pachydrobiini, in China, 181
Pachydrobiini Davis and Kang, 163–164
Parallelism, 185
Paromomycin, 13
Passive immunization, antibody-mediated, 112–113
Phlegmonous inflammation, 73
Phlegmonous stage, 74, 79
Phocanema, occurrence in world, 44
Phylogenetic analysis, characters and character-states used for, 170–171
Plasmodium, vi, 103–107
  life cycle of, 104
Plasmodium berghei sporozoites, 105
Plasmodium falciparum, human vaccine trials and, 106–107
Plasmodium sporozoites, 103
Plasmodium vivax sporozoites, 110
Plasmodium yoelii sporozoites, 109
Pomatiopsidae family, 154
Pomatiopsinae, 146, 161–163
  distribution of genera geographically, 176
  divergence of, 150

Pomatiopsinae subfamily, 154
Pomatiopsine genera, 152
Pomatiopsini, 161–162
Pregnancy, toxoplasmosis and, 25–26
Promastigotes, 119
Prosobranch snails, vii, 145–200
  adaptive and morphostatic radiations, 195–196
  classification, 153–165
  coevolution, 182–194
  coevolved specificity involving, 187–192
  ecology of transmission, 195–200
  general habitats of, 197
  higher taxa defined, 154–161
  historical reconstruction, 192–194
  hypotheses concerning, 147–153
  morphology and clades, 165–178
  old problems resolved, 178–182
Pseudobythinellini, 162–163
  shells of, 156–157
Pseudoterranova, genera of, 49–50
Pseudoterranova decipiens, 44
  life cycle of, 54–56
  third-stage larvae, 55
Pseudoterranovosis
  gastric, 45–46
  intestinal, 46
Pyrimethamine, 33, 35–36

R
Renette cells, 73
Restriction fragment length polymorphisms (RFLPs), 90, 92–93
Rivers, major, of Asia, 149
Robertsiella, 198
  shells of, 155
Roxithromycin, 36

S
Sake, 64–65
Sashimi, 64–65
Schistosoma
  cross-infectivity studies, 191
  divergence of species of, 189
  hypotheses concerning, 147–153
  infectivity studies, 194
  number of species transmitted by triculines, 188
  Prosobranch snails and, vii, 145–200
Schistosoma japonicum, 146

Schistosoma japonicum complex, 182–183
Schistosoma mekongi, 200
Schistosoma sinensium complex species, 186–187
Schizonts, 3, 5, 14, 15
Shoga, 64–65
Skip lesions, 71
Snails, Prosobranch, *see* Prosobranch snails
Spiramycin, 12
Sporozoite immunity, 104–105
Sporozoite vaccines, 107
Sporozoites, 3
  malaria, vi, 103–107
  Plasmodium, 103
  Plasmodium berghei, 105
  Plasmodium vivax, 110
  Plasmodium yoelli, 109
Stomach, vanishing tumor of, 56–61
Submucosal layer, 73
Sulfadiazine, 33, 35–36
Synapomorphies, 170

T

T-cell responses in leishmaniasis, vi–vii, 119–137
T-cell subsets in human leishmaniasis, 125–128
T-cells, 123–124
  human, *Leishmania* antigens stimulating, 128–132
Tachyzoites, 22
Terranovasis, in Japan, 46
Thysanoessae, 49
Tissue cyst rupture, 25
Tissue cysts, 22, 23
Toxoplasma gondii, 22–25
Toxoplasmic encephalitis, 21, 31, 33–36
Toxoplasmic lymphadenitis, 31
Toxoplasmosis, v–vi, 21–37
  AIDS and, 27, 28
  clinical presentation, 25–28
  congenital, 26
  diagnosis, 28–33
  epidemiology and transmission, 23
  ocular, 26
  organism, 22–25
  pathogenesis and host defenses, 24–25
  pregnancy and, 25–26
  treatment, 33, 35–37
Traveler's diarrhea, 6
Tricula, 146
Tricula bollingi, 151
Triculinae, 163–165
  differentiating genera of, 166–167
  divergence of, 150
  shells of, 156–157
Triculinae adaptive radiation, 195–196
Triculinae subfamily, 154
Triculine genera, 152
Triculine phylogenies, 145–153
Triculine shells, 158–160
Triculine species diversity, localities rich in, 172
Triculines, number of species of *Schistosoma* transmitted by, 188
Triculini, in China, 181
Triculini Bayis, 164
Trimetrexate, 36
Trophozoites, 3, 22

V

Vaccines
  malarial, vi, 103–114
  Plasmodium falciparum and human trials of, 106–107
  sporozoite, 107
  subunit, antibody induction with, 108–109
Vanishing tumor of stomach, 56–61
Visceral leishmaniasis, acute (VL), 120

W

Wasabi, 64–65

X

D-Xylose test, 8

Y

Yeasts, 10

Z

Zidovudine (AZT), 12

MIX
Papier aus verantwortungsvollen Quellen
Paper from responsible sources
FSC® C105338

If you have any concerns about our products,
you can contact us on
ProductSafety@springernature.com

In case Publisher is established outside the EU,
the EU authorized representative is:
**Springer Nature Customer Service Center GmbH
Europaplatz 3, 69115 Heidelberg, Germany**

Printed by Libri Plureos GmbH
in Hamburg, Germany